Imagining the Future of Climate Change

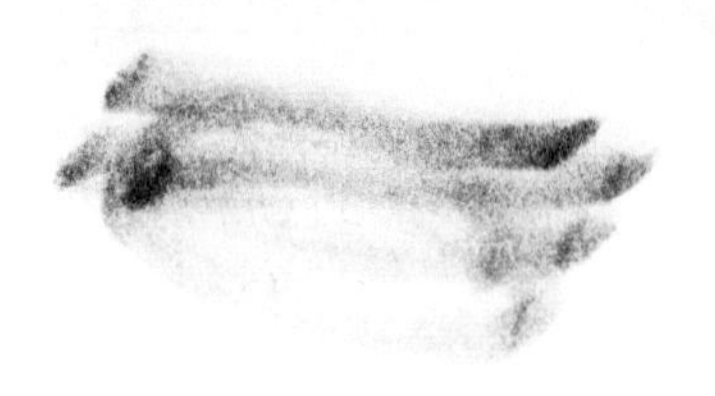

AMERICAN STUDIES NOW:
CRITICAL HISTORIES OF THE PRESENT

Edited by Lisa Duggan and Curtis Marez

Much of the most exciting contemporary work in American Studies refuses the distinction between politics and culture, focusing on historical cultures of power and protest on the one hand, or the political meanings and consequences of cultural practices, on the other. American Studies Now offers concise, accessible, authoritative, e-first books on significant political debates, personalities, and popular cultural phenomena quickly, while such teachable moments are at the forefront of public consciousness.

1. *We Demand: The University and Student Protests,* by Roderick A. Ferguson
2. *The Fifty-Year Rebellion: How the U.S. Political Crisis Began in Detroit,* by Scott Kurashige
3. *Trans*: A Quick and Quirky Account of Gender Variability,* by Jack Halberstam
4. *Boycott! The Academy and Justice for Palestine,* by Sunaina Maira
5. *Imagining the Future of Climate Change: World-Making through Science Fiction and Activism,* by Shelley Streeby

Imagining the Future of Climate Change

World-Making through Science Fiction and Activism

Shelley Streeby

UNIVERSITY OF CALIFORNIA PRESS

University of California Press, one of the most distinguished university presses in the United States, enriches lives around the world by advancing scholarship in the humanities, social sciences, and natural sciences. Its activities are supported by the UC Press Foundation and by philanthropic contributions from individuals and institutions. For more information, visit www.ucpress.edu.

University of California Press
Oakland, California

Library of Congress Cataloging-in-Publication Data

Names: Streeby, Shelley, 1963– author.
Title: Imagining the future of climate change : world-making through science fiction and activism / Shelley Streeby.
Description: Oakland, California : University of California Press, [2018] | Includes bibliographical references. | Identifiers: LCCN 2017034941 (print) | LCCN 2017039458 (ebook) | ISBN 9780520967557 (epub) | ISBN 9780520294448 (cloth : alk. paper) | ISBN 9780520294455 (pbk. : alk. paper)
Subjects: LCSH: Climatic changes. | Global warming. | Indigenous peoples—Ecology—United States.
Classification: LCC QC902.9 (ebook) | LCC QC902.9 .S77 2018 (print) | DDC 304.2/80897—dc23
LC record available at https://lccn.loc.gov/2017034941

Manufactured in the United States of America

26 25 24 23 22 21 20 19 18
10 9 8 7 6 5 4 3 2 1

For Curtis, for talking to me every day, reading everything I write, writing his own beautiful book on farmworker futurisms, and being the best companion I can possibly imagine in living a life that I love

For Ayana, for collaborating with me, doing things for the right reasons, and creating a community of writers, scholars, artists, and activists keeping alive Octavia's memory

For all the Clarion students and instructors I've met since I became director in 2010, for teaching and inspiring me all these years

CONTENTS

Overview ix

Introduction
Imagining the Future of Climate Change
1

1. #NoDAPL
Native American and Indigenous Science, Fiction, and Futurisms
34

2. Climate Refugees in the Greenhouse World
Archiving Global Warming with Octavia E. Butler
69

3. Climate Change as a World Problem
Shaping Change in the Wake of Disaster
101

Acknowledgments 127

Notes 131

Glossary 147

Key Figures 151

Selected Bibliography 155

OVERVIEW

INTRODUCTION: IMAGINING THE FUTURE OF CLIMATE CHANGE

Indigenous people and people of color use science fiction and futurisms to make movements at the forefront of imagining the future of climate change.

Snowpiercer · *Geo-engineering* · *Global Warming* · *Cli-fi* · *Science Fiction* · *Speculative Fiction* · *Visionary Fiction* · *Social Movements*

CHAPTER 1. #NODAPL

Indigenous science, fiction, and futurisms shaped the #NoDAPL struggle led by the Standing Rock Sioux tribe, as well as other worldwide struggles over oil, water, and resource extraction, including in Māori contexts. Indigenous-helmed movements practice world-making through taking direct action, working in Indigenous science and technologies, and imagining decolonized futures in the wake of climate change disaster in many different kinds of speculative fiction across multiple media platforms.

"Let Our Indigenous Voices Be Heard" · *Indigenous Futurisms* · *Direct Action* · *#NoDAPL* · *Standing Rock* · *Mni Wiconi (Lakota for "Water Is Life")* · *Native Slipstream* · *Indigenous Environmental Network* · *Leslie Marmon Silko,* Almanac of the Dead · Anamata Future News

CHAPTER 2. CLIMATE REFUGEES IN THE GREENHOUSE WORLD

The great science fiction writer Octavia E. Butler was a major climate change intellectual whose extrapolations from her present, theorizing of climate refugees, and archival memory-work illuminate blind spots in 1970s to early 2000s climate change conversations. Butler imagined symbiotic possibilities for shaping change in a world transformed by the greenhouse effect.

Climate Refugees · *Archive* · *HistoFuturist* · *Neoliberalism* · *Symbiosis* · *Extrapolation* · *Greenhouse Effect* · *Disaster* · *Critical Dystopia* · Parable of the Sower

CHAPTER 3. CLIMATE CHANGE AS A WORLD PROBLEM

From the early 2000s to the present, Detroit organizer, theorist, and fiction writer adrienne maree brown has created generative intersectional connections among Octavia E. Butler's work, science fiction, social movements, and direct action. Brown's theorizing of emergent strategy, crafting of visionary fiction, and learning from and connecting to Indigenous people are especially significant contributions to imagining the future of climate justice as a world problem.

U.S. Social Forum · *Allied Media Conference* · *Detroit* · *Bali Principles of Climate Justice* · *Inuit Circumpolar Council* · *UN Declaration on the Rights of Indigenous Peoples* · *Arctic Indigenous Youth Alliance* · *Direct Action* · *Grace Lee Boggs* · *Ruckus Society* · Octavia's Brood: Science Fiction Stories from Social Justice Movements

Introduction

Imagining the Future of Climate Change

In Bong Joon-Ho's 2013 international blockbuster *Snowpiercer,* a single train traverses the globe, protecting the final remnant of humanity from an Ice Age that made the planet seemingly uninhabitable. Set in 2031, the film begins by squarely placing the blame on humans for this catastrophic climate change. As the opening credits roll over dark, starry space, we hear crackling, fuzzy excerpts of news broadcasts from all over the world telling how, despite protests from environmental groups and "developing countries," on 1 July 2014, seventy countries dispersed the artificial cooling substance CW-7 into the upper layers of the atmosphere. Because "global warming can no longer be ignored," one of the disembodied voices explains, seeding the skies with CW-7 was a last-ditch effort "to bring average global temperatures down to manageable levels as a revolutionary solution to mankind's warming of the planet." But before the real action of the film even starts, we learn the grim outcome of this desperate international scientific experiment. Two short sentences loom large on the screen: "Soon after dispersing CW-7 the world froze. All life became extinct."

In imagining a geo-engineering experiment gone wrong in response to the disaster of global warming, Bong is asking us to think critically about the solution to the problem of climate change that is favored by many people, states, and corporations invested in finding alternatives to curbing carbon emissions.[1] In addition, since the mid-2000s, some prominent scientists, such as Dutch Nobel Prize-winning atmospheric chemist Paul Crutzen, have also advocated exploring geo-engineering out of fears that states will not take the necessary actions to curb global warming in time and that we are on the brink of locking in dystopian climate change that will render unsustainable life on Earth as we know it. Geo-engineering refers, in Clive Hamilton's words, to "deliberate, large-scale intervention in the climate system designed to counter global warming or offset some of its effects."[2] Hamilton distinguishes two classes of geo-engineering: carbon dioxide removal technologies that try to remove excess carbon dioxide from the atmosphere and store it elsewhere, and solar radiation management technologies, which aim to reduce the amount of sunlight reaching the planet, thereby mitigating one of the most prominent symptoms of global warming without fixing the cause. Many techno-fix fantasies imagine blocking the sun through a range of methods including space mirrors, spraying seawater into the sky to create more cloud cover, or spraying sulfate aerosols into the stratosphere, as we see in *Snowpiercer.* According to Hamilton, "the idea of spraying sulfite particles into the upper atmosphere was sparked by observing the effects on the weather of volcanic eruptions"—a phenomenon that scientists have been aware of as far back as the eighteenth century—which provoked scientists to imagine "countering global warming by mimicking the cooling effect of volcanoes" (59). Hamilton calls stratospheric aerosol spraying "the archetypal geo-engineering technique" since "it would be easy, effective, and cheap, and have the most far-reaching implications for life on Earth" (59). Geo-engineering projects carry significant risks, however, since as Hamilton puts it, the earth's climate is a nonlinear, complex system and introducing changes may create unpredictable effects, including, among others in the case of aerosol spraying, the possibility of disrupting the Indian monsoon, thereby "affecting food supplies for up to two billion people" (64).

In interviews, Bong clarifies that he was indeed thinking of geo-engineering as a hubristic project that introduces giant risks for huge parts of the world in an effort to keep the machine of global fossil fuel capitalism going. In a press kit released with the film, the synopsis also emphasizes the connection between climate change and class inequalities: "Climate change has made the planet uninhabitable" and "the world inside the train is far from equal."[3] When asked if the film is a response to climate change, Bong replied that while in South Korea people talk about how China's environmental issues impact Korea and circulate rumors about China's geo-engineering projects, he was trying to call attention to "how big business tries to both use and control nature," since "it's not in their interests to change." He also claims "it's not humans per se, but capitalism that's destroying the environment" and that if we could "control human greed," it would "go a long way towards slowing down our ongoing environmental disaster."[4]

As the recent proliferation of geo-engineering schemes suggests, the idea that humans can master nature without risk or cost is a deep fantasy, but in *Snowpiercer,* as in many such attempts to control nature in the history of speculative fiction, arguably beginning with Mary Shelley's *Frankenstein,* this effort backfires. In Bong's words, "Nature takes its revenge and sends them back to the ice age."[5] Bong further explains that *Snowpiercer* is a science fiction film precisely because the latter is "a genre where you can express the human condition and systems in which we live much more directly and symbolically," which helped him explore questions about climate change and global class inequalities and stage them for a global audience.[6]

Snowpiercer is only one of many recent speculative fictions that make climate change the central problem in imagining the future, often in a dystopian mode. That's not surprising, because imagining the future of climate change at this moment is frightening. For years now scientists have issued warnings about what will happen if we fail to act soon. More dramatic and destructive storms, the loss of biodiversity, species extinction, and sea level rise are just a few of the changes that are no longer on the horizon but are happening now. Every day, new stories circulate about the latest signs of impending catastrophic climate change. Meanwhile, radically transformed climates are at the heart of a lot of science fiction, so much so that a whole new subgenre called cli-fi has emerged. Cli-fi or climate change fiction is best situated within the larger category of speculative fiction, an umbrella genre that includes science fiction and fantasy. In 2013, National Public Radio (NPR) and the *Christian Science Monitor* began to use the term cli-fi to encompass a wide variety of dystopian visions of near-future climate change, including Barbara Kingsolver's *Flight Behavior,* Nathaniel Rich's *Odds against Tomorrow,* and Margaret Atwood's *The Year of the Flood.* Since then the subgenre has exploded.[7]

While I dip into cli-fi here and there in this book, in what follows I tell the story of imagining the future of climate change by focusing especially on movements, speculative fictions, and futurisms of Indigenous people and people of color—work that is all too often excluded from the category of cli-fi and that extends beyond cli-fi in its rich and deep connections to social movements and everyday struggles and to other cultural forms such as film, video, music, social media, and performance. In Amitav Ghosh's bracing book *The Great Derangement: Climate Change and the Unthinkable,* he, like many before him, excludes science fiction from serious consideration as a contributor to debates over climate change, arguing, following Margaret Atwood, that "the Anthropocene resists science fiction" because the latter focuses on "an imagined other world located apart from our ours." He also argues that despite a few notable exceptions such as Liz Jensen's and Barbara Kingsolver's novels, even cli-fi, with its realist elements, fails because it "is made up of disaster stories set in the future" rather than examining the recent past and present.[8] In contrast, I argue in what follows that people of color and Indigenous people use science fiction and other speculative genres to remember the past and imagine futures that help us think critically about the present and connect climate change to social movements.

Here and throughout this book I distinguish between people of color and Indigenous people even though historically these identities often intersect and converge. I make this distinction in order to recognize particular histories of settler colonialism, treaty-making, dispossession, nationhood, and citizenship that situate Natives differently than non-Native people of color in the United States and the Americas. Settler colonialism is a distinct kind of colonialism that aims to eliminate and replace Natives by settling on and extracting value from their lands.[9] Furthermore, since 1924, Native Americans have possessed dual citizenship: they are documented as citizens by their tribal nations as well as by the United States. The use of the term "people of color" in the United States, on the other hand, can be traced at least as far back as the French colonies in the Americas, where it was used to refer to people of mixed African and European descent who were not slaves. It is currently a keyword in scholarship on race and ethnicity in the fields of ethnic studies and American studies, where it refers to people who are not white. Often, such scholarship explores coalitions, solidarities, and social movements that connect groups, while also attending to contradictions and differences that shape the latters' relations to each other, the United States, and the world. In that spirit, in what follows I analyze how Indigenous people and people of color in the United States, through their art, activism, and speculative fictions, respond to climate change by imagining futures that are sometimes in sync with each other and sometimes not. Although this is a selective lens for envisioning the future of climate change, it is a richly illuminating one that yields important insights and possibilities that we miss when the focus is only on nation-states, transnational corporations, research scientists, and politicians as significant agents and explainers of change.

In focusing on social movements and cultures of climate change, I build on "social movements and culture" methodologies used in American Studies. As modeled by scholars such as Michael Denning and George Lipsitz, such methodologies look for meaning in the connections people make between cultural texts and the important social movements of their times. Today a transnational movement from below, significantly led by Indigenous people and people of color, is one of the most powerful forces opposing the fossil fuel industry's transnationalism from above. My goal is to introduce the history and most significant flashpoints in imagining the future of climate change over which these movements currently struggle. Speculative fiction and

Indigenous and people of color futurisms both illuminate and make that history. But first it is necessary to understand the theory of global warming that is also central to that history.

A BRIEF HISTORY OF GLOBAL WARMING

Earth's temperature is determined by the difference between the energy received from the sun and the amount that is released back into space. Ozone absorbs some incoming solar shortwave radiation and about a third of the solar energy returns to space, while the land and ocean absorb what's left. The land and ocean then radiate this warmth "as long-wave infrared or 'heat radiation.' Atmospheric gases such as water vapor, carbon dioxide, methane, and nitrous oxide are known as greenhouse gases as they can absorb some of this long-wave radiation, thus warming the atmosphere." This is what we call the "greenhouse effect": "Since, the industrial revolution we have been burning fossil fuels (oil, coal, natural gas) deposited hundreds of millions years ago, releasing the carbon back into the atmosphere as CO2 and CH4, increasing the 'greenhouse effect' and elevating the temperature of the Earth." Within the span of one century, we have put more carbon into the atmosphere than during the previous thousands of years.[10]

Mathematician Joseph Fourier first formulated what we now call the theory of the greenhouse effect in 1827. Three decades later, in 1859, John Tyndall identified carbon dioxide, methane, and water vapor as greenhouse gases, and in 1896 Svante Arrhenius made remarkably astute predictions of how much the climate would change in response to changing concentrations of carbon dioxide in the atmosphere. He calculated that doubling carbon dioxide would increase the temperature of Earth by an average of 4 to 6 degrees Celsius, and, according to David

Archer and Stefan Rahmstorf, "in spite of the crudeness of the data available and a few questionable assumptions, Arrhenius got the answer basically correct."[11] During the 1940s, technologies of measuring CO_2 radiation interception improved dramatically and in 1955, Gilbert Plass proved that adding CO_2 to the atmosphere intercepted more infrared radiation and kept it from being lost to space, thereby warming the planet. Finally, at the end of the decade in 1959, Plass published an article in the *Scientific American* called "Carbon Dioxide and Climate," in which he ominously warned that "if carbon dioxide is the most important factor" in increasing Earth's temperature," then "long-term temperature records will rise continuously as long as man consumes the earth's reserves of fossil fuels."[12]

Still many scientists, including Plass himself, believed oceans might serve as giant sinks absorbing the extra carbon dioxide produced by humans until Roger Revelle and Hans Suess of the Scripps Institute of Oceanography in La Jolla, California, challenged that idea by arguing that sea water was already saturated with carbon dioxide and thus oceans would not be able to absorb the excess produced by humans to the extent previously imagined; they ominously concluded that carbon dioxide was therefore very likely increasing in the atmosphere. In 1958, Charles David Keeling began taking daily measurements of the concentration of atmospheric carbon dioxide at the Mauna Loa Observatory, a project that has continued up to this day. As a result, he devised what is now called the Keeling CO_2 curve, a graph that plots the ongoing change in concentrations of carbon dioxide in Earth's atmosphere since 1958. This evidence helped Keeling demonstrate the existence of a cycle that responded not only to the growth and decay of land plants in the northern hemisphere but also to long-term increases created by burning fossil fuels.

Within "a very few years he could see that the annual maximum value for CO_2 was steadily rising."[13]

In response to new research as well as the concern for the environment sparked by Rachel Carson's book *Silent Spring* (1962), the 1960s witnessed the proliferation of grassroots movements, large nonprofit organizations, and environmental institutions created by nation-states. One important flashpoint was the formation of a U.S. President's Science Advisory Committee on Environmental Pollution, which in 1965 announced that "pollutants have altered on a global scale the carbon dioxide content of the air."[14] An appendix entitled "Atmospheric Carbon Dioxide" partly authored by Keeling and Revelle explained in detail how carbon dioxide that remains in the atmosphere has "significant effect on climate," acting "much like the glass in a greenhouse" to "warm the temperature of the lower air" (113). The authors warned that "through his worldwide industrial civilization, Man is unwittingly conducting a vast geophysical experiment" (126), burning within a few centuries the carbon that had accumulated for the last five hundred million years, and predicted the possibility of the melting of the Antarctic ice cap, catastrophic sea level rise, the warming of ocean waters, and many other disasters if nothing was done. Another important organization formed in 1967 was the Environmental Defense Fund, a U.S.-based nonprofit created by scientists as part of an effort to ban DDT but that grew into a major environmental advocacy group, albeit one that is now widely criticized for its collaborations with big corporations and business-friendly solutions to environmental problems, something that continues to limit the effectiveness of the mainstream environmental movement today.

In the 1970s, many scientific research projects focusing on carbon dioxide and climate emerged to build on Keeling's and

Revelle's work, while the dramatic impact of human release of other greenhouse gases such as methane, chlorofluorocarbons, and nitrous oxide was also measured. Next, a flurry of government institutions was created in response to emerging public concern and pressure about environmental problems. Following the 1969 National Environmental Policy Act (NEPA), several new environmental laws were passed, including one requiring environmental impact reports for major state projects. The next year, in 1970, the first Earth Day took place in the United States. Earth Day was the idea of Wisconsin Senator Gaylord Nelson, who proposed a massive teach-in after witnessing the ravages of the 1969 massive oil spill in Santa Barbara, California. Twenty million people ended up participating in these events, which took place across the nation, received wide media coverage, and precipitated the formation that year of the Environmental Protection Agency and the passage of the Clean Air, Clean Water, and Endangered Species Acts. Also in 1970, the U.S. National Oceanic and Atmospheric Administration, which would become the world's leading funder of climate research, was formed. While the first UN Environmental Conference in Stockholm in 1972 devoted little time to climate change, in 1975, the coinage by U.S. scientist Wallace Broecker of the term "global warming" in a scientific paper introduced the phrase into the language of science and eventually into official reports and media stories.

In an article called "Climatic Change: Are We on the Brink of a Pronounced Global Warming?" in *Science* magazine, Broecker warned that after the next decade "the CO_2 effect will tend to become a significant factor and by the first decade of the next century we may experience global temperatures warmer than any in the last one thousand years"—which in fact has proven to be the case.[15] Then, in 1978, President Carter's decision to resort to U.S.-produced coal in the face of the oil embargo put carbon dioxide production squarely on the political map. By 1979, the National Research Council declared there was now "incontrovertible evidence that the atmosphere is indeed changing and that we ourselves contribute to that change" as well as a "consensus" that there will be a "warmer earth with a different distribution of climatic regimes." In order to adequately address the question of how these changes would affect the complex web of life, the authors of the report noted, one would have to "peer into the world of our grandchildren."[16] Although the report did not go that far, leaving it to creators of speculative fiction to imagine future worlds transformed by climate change, it did warn that waiting to see might mean waiting too late.

While changing weather, warming oceans, and warnings about the future of the world's ecology and environment made the news as early as the 1960s, the 1980s was the key decade when climate change became a central topic in the media. By the 1980s, the global mean temperature was increasing rapidly, with 1981 the warmest year on record, while developments in climate modeling and research on climate history revealed how quickly transitions to warmer periods could happen, partly due to feedback loops that kick in when ice sheets start to melt, such as sea level rise. The election in 1980 of Ronald Reagan as president of the United States proved a serious setback to the emerging environmental movement since his administration was hostile to the small gains that had been made, pushed deregulation, and prioritized untrammeled economic growth over confronting environmental harms. In 1983, the National Academy of Sciences issued a new report which stated that carbon dioxide in the last generation had increased from 315 to 340 parts per million by volume and that this increase was primarily attributable to burning of coal, oil, and gases created by human activity. They concluded that as a result global mean temperatures would continue to rise, which would significantly reduce the availability of water in places such as the U.S Southwest and also threatened to cause dramatic sea level rises and the eventual disappearance of the West Antarctic Ice Sheet.[17] In 1985, the climate change alarm was sounded again by the British Antarctic Survey's report of ozone depletion over Antarctica and by 1987, the Vienna Convention's Montreal Protocol set international limits on the emission of gases that adversely affected the ozone.

In 1988 serious discussion of the need to reduce greenhouse gas emissions began to emerge as news coverage of global warming dramatically increased following a year of heat waves and droughts. It was also the year that scientist James Hansen testified before the U.S. Senate that "Global Warming Has Begun," as a 24 June *New York Times* headline put it. Hansen made history by telling the room of politicians that '"It is time to stop waffling, and say that the evidence is pretty strong that the greenhouse effect is here." That same year, the United Nations Environment Program (UNEP) and the World Meteorological Organization (WMO) established the Intergovernmental Panel on Climate Change (IPCC) to survey research on climate change and its potential environmental and socioeconomic impacts.

One year later, in 1989, in response to all this environmentalist activity and pressure, fossil fuel interests created the Global Climate Coalition (GCC), a group whose main focus was introducing doubt in the minds of citizens and politicians about the validity of climate science—a project that vigorously persists today despite the formal demise of this particular group in the early 2000s.[18] In a 2009 *New York Times* story, journalist Andrew Revkin reported on how for over a decade the GCC, which represented industries whose profits depended on fossil fuels, "led an aggressive lobbying and public relations campaign against the idea that emissions of heat-trapping gases could lead to global warming."[19] Although internal documents later revealed that the organization's scientists largely agreed with the emerging consensus that burning fossil fuels was the biggest contributor to global warming, they nonetheless put huge amounts of money into arguing against the idea that international agreements were necessary in response, especially after the "Earth Summit," the 1992 Rio de Janeiro United Nations Conference on Environment and Development (UNCED) that took place in June 1992. For instance, the group spent 1.6 million dollars in 1997 alone, the year of the Kyoto Protocol, the international treaty in which 193 states agreed to reduce greenhouse gas emissions. According to climate scientist Benjamin Santer, IPCC author and one of the targets of the group's ire, the coalition was "'engaging in a full-court press at the time, trying to cast doubt on the bottom-line conclusion of the I.P.C.C.,' which had concluded in 1995 that 'the balance of evidence suggests a discernible human influence on global climate.'"[20] The 1991 and 1995 IPCC reports warned that burning fossil fuels was raising the mean global temperature of the planet and predicted significant sea level rise and other harms would cause catastrophic social, economic, and political problems if nothing was done to mitigate current patterns of greenhouse gas emissions. The 1997 Kyoto Protocol represented one response to increasing public awareness of this scientific research, as people pressed governments to act. Although the United States never ratified the protocol and though throughout the nineties the GCC continued to spend large amounts of money to undermine climate change science, reports of the breaking up of Antarctic ice sheets and signs of warming in polar regions continued to shape public opinion and make the environment and climate an object of concern for social movements in the United States and around the world.

From the 1960s through the 1990s, environmental movements that emerged in response to climate change were the single biggest forces pressing states to act, while fossil fuel industry advocates lobbied hard against regulations, spreading doubt about climate change science, even as big polluter states like the United States and China balked at bigger changes. Critical to the formation of such social movements were cultural texts that moved large numbers of people to act and imagine alternatives to the greenhouse fossil fuel world. One of the earliest and most important of these was Rachel Carson's book *Silent Spring* (1962), which sold over two million copies and was translated into at least seventeen languages.[21] Carson was an aquatic biologist who became a nature writer in the 1950s. After writing three books about ocean life, she became financially independent and decided to write a book on how pesticides were altering human bodies and the planet. She uncovered a vast amount of evidence for the pesticide-cancer connection and also confronted the problem that persists to this day of industry experts putting resources into covering up or denying problems instead of addressing and solving them.

The book makes a forceful case for human-created damage to the planet and asks readers to consider how so-called scientific advances may also create new problems. It was incredibly successful at calling attention to such dangers and has been credited with provoking the creation of the Environmental Protection Agency by the Nixon Administration in 1970 and the phasing out of DDT by 1972 as activists and environmental organizations emerged to push forward Carson's research. The great science fiction writer James Tiptree, Jr. / Alice Sheldon, a gender-bending woman who wrote under a male pseudonym for most of her life, looked back in the mid-70s and commented that, though most women, like the mythical Cassandra, are doomed to speak the truth and never to be believed, Carson was "maybe the last to break through" with "an unpleasant truth" and still be heard.[22] Despite being attacked by the chemical companies, Carson's research held up and she was widely recognized as a hero of the environment, although sadly it was soon discovered that she had breast cancer. She was weakened by radiation treatments as the acclaim and criticisms began to roll in, and though she continued to make appearances to support *Silent Spring*'s findings, she died less than two years after the publication of her world-changing book. In her wake, *Silent Spring* inspired a movement.

SPECULATIVE FICTIONS OF CLIMATE CHANGE

Silent Spring changed the world because of the way it was written and also because of Carson's creative use of multiple media platforms to communicate her message. The book was initially serialized in three parts in three June 1962 issues of the *New Yorker* magazine, where it caught the eye of many readers and the chemical industry. Excerpts were also serialized in *Audubon* magazine and the *New York Times* published a positive editorial about it. One of the biggest boosts to *Silent Spring*'s popularity came when it was chosen as a Book of the Month Club selection for October, which Carson observed made people aware of it in parts of the United States that "didn't know what a bookstore looks like—let alone *The New Yorker*."[23] Her appearance on the TV show *CBS Reports* on April 3, 1963, entitled "The Silent Spring of Rachel Carson," also made a big impact on the new TV-watching public, as Carson went head to head with Robert White-Stevens, a spokesman for the agricultural chemical industry, and calmly bested him with her steady demeanor and mastery of facts. As the *New York Times* put it the day after the program aired, after watching it a "lay viewer" had been exposed to the logic of both sides and "would still agree with the program's central conclusion": that Carson's "theme is sufficiently persuasive and disturbing" to warrant "intensified research" on "the long-term consequences" of pesticides and other chemicals.[24] Her testimony before the Senate Commerce Committee in June 1963 was also widely covered in the media.

Notably, in order to engage the nonscientist reader, *Silent Spring* begins with a piece of speculative fiction, "A Fable for Tomorrow." This brief chapter focuses on an imaginary town that "does not actually exist" but "might easily have a thousand counterparts in America or elsewhere in the world" (3). While the narrator initially invites the reader to picture "the abundance and variety" of the town's "bird life" (2) and its clear, cold water full of fish and other living things, the tone suddenly, dramatically shifts as a "strange blight" falls over the earth, causing animals to die, introducing new diseases into families, producing unexplained deaths, even among children, and making the birds disappear. No bees come to pollinate and the fish disappear from the waters as this "stricken world" of endless "disasters" (3) transforms into a harsh, forbidding, dystopian landscape.

Imagining the silencing of "the voices of spring" in "countless towns" (3) across the nation, Carson's speculative fiction was so effective that it provoked a response in kind from Monsanto, which produced its own corporate speculative fiction, called "The Desolate Year." Published in *Monsanto Magazine* and widely distributed, it asked readers to imagine a year without pesticides, warning that disgusting, rapacious bugs would invade the world and turn up everywhere, including "yes—inside man," who would have "no weapon but a fly swatter against rampant bed bugs, silverfish, fleas, slithering cockroaches, and spreading ants," that ticks would "leap" onto "people," and slice "deeper and deeper" into their flesh, gorging on it and becoming "many times their normal size.[25] Despite such corporate efforts, *Silent Spring's* popularity led to the banning of eight of the twelve pesticides covered in her book, while restrictions were put on three others as well.

Two of the most enduring and important legacies of Carson's book, however, are modern environmental movements and the continuing struggle over the role of states in defending the environment and the planet. In response to Carson's work and to DDT tragedies on Long Island, in 1967 ten scientists formed the Environmental Defense Fund, which began bringing lawsuits against the state to establish citizens' rights to a clean environment. The state responded with a flurry of new policies and institutions, including the passing, in 1969, of the National Environmental Policy Act (NEPA), which created a Council on Environmental Quality (CEQ) at the White House and introduced a new requirement for environmental impact statements. In December 1970, the U.S. Environmental Protection Agency (EPA) was created. In the more than half century that has passed since *Silent Spring's* publication, Carson has become a hero of the environment for a new generation of activists who are inspired by her example. This impact crucially depended on her imaginative approach to communicating science to lay readers and her talent at using multiple media platforms and cultural forms to make people care about stopping the "reckless and irresponsible poisoning of the world that man shares with all other creatures" and learning, instead, to cultivate "our accommodation to the world that surrounds us" (ix).

Today's cli-fi writers are connected to Carson and indebted to her as a precursor in their efforts to combine science and speculative fiction, which are evident in recent anthologies such as *Loosed Upon the World: The Saga Anthology of Climate Fiction* (2015). This collection includes several climate change stories by notable writers such as Kim Stanley Robinson, Atwood, and Paolo Bacigalupi, author of many award-winning stories and novels including *The Water Knife* (2015), a near-future thriller about climate change, drought, and water wars set in the U.S. Southwest. In a foreword, Bacigalupi argues that when creating a cli-fi world, "you have the chance of making people engage not with the future, but with the intense realities of our present—the realities that were previously passing them by." He hopes that by experiencing climate change "viscerally" through fiction, instead of abstractly or theoretically, readers of cli-fi will be ready to "think long term effectively."[26] Bacigalupi's insight that climate change fiction encourages people to engage with "intense realities" of the present that might otherwise go unnoticed resonates with the great writer and critic Samuel Delany's remark that "science fiction is not about the future; it uses the future as a narrative convention to present significant distortions of the present." In other words, Delany suggests that some science fiction stories can help us take hold of the present and engage its intense realities instead of passively letting it pass by without thinking about where things are heading.[27] A good deal of cli-fi works on this principle, distorting our present by representing it as the past of an imagined future, as literary critic Fredric Jameson says classic science fiction writer Philip K. Dick often does, in ways that can help us think critically about what we need to do in the present to keep the worst from happening.[28] But because cli-fi is a capacious subgenre that incorporates other elements besides science fiction, the umbrella term "speculative fiction" is a more useful category within which to place it.

The term "speculative fiction" has a long and complicated history that, like the conversation about human-caused climate change, goes back at least to the nineteenth century. In 1889 *Lippincott's Monthly Magazine* used the term to describe Edward Bellamy's utopian novel *Looking Backwards, 2000–1887* and several other fictions set "in the future tense." Robert Heinlein, author of dozens of science fiction novels, notably including *Starship Troopers* (1959) and *Stranger in a Strange Land* (1961), also used the term in the mid-twentieth century, at first as interchangeable with "science fiction" and in ways that specifically excluded fantasy. In 1947 Heinlein wrote that he preferred "speculative fiction" to "science fiction" because the term better captured the genre's ability to ask big and important questions about "sociology, psychology, esoteric aspects of biology, impact of terrestrial culture on the other cultures we may encounter when we conquer space, etc., without end." Heinlein insisted, however, that speculative fiction "is not fantasy fiction, as it rules out the use of anything as material which violates established scientific fact, laws of nature, call it what you will, i.e. it must be possible to the universe as we know it."[29] In the 1960s, New Wave anthologist and writer Judith Merril used the term in the subtitle of *England Swings SF: Stories of Speculative Fiction* (1968) to distinguish the New Wave from earlier pulp science fiction, though New Wave politics were generally well to the left of Heinlein. In the introduction, Merril calls the book a "collection of science fiction, social criticism, surrealism ... what have you" and promises readers a "good trip."

Margaret Atwood has used the term "speculative fiction" to classify her famous novel *The Handmaid's Tale* (1985) because she says that, though the novel is set in the future, it projects from trends "which are already in motion." She contrasts speculative fiction with science fiction, which she defines as "fiction in which things happen that are not possible today—that depend for instance on advanced space travel, time travel, the discovery of green monsters on other planets or galaxies, or that contain various technologies we have not yet developed."[30] Ursula K. Le Guin, among others, has criticized Atwood's "arbitrarily restrictive definition," which she speculates "seems designed to protect her novels from being relegated to a genre still shunned by hidebound readers, reviewers and prize-awarders." Le Guin claims, on the other hand, that "one of the things" science fiction does is "extrapolate imaginatively from current trends and events to a near-future that's half prediction, half satire," as she believes most of Atwood's novels do.[31] Clearly the use of the term "speculative fiction" has sometimes been a distinction-forging move aimed at rescuing the genre from disparagement by providing a more respectable genealogy, as Heinlein and Atwood did, or by making it a more experimental, boundary-crossing, transgressive force connected to social criticism and surrealism, as Merril exuberantly imagined it. But "science fiction" also has a specific history and constellation of meanings worth remembering in a book on imagining the future of climate change. In what follows, I use both terms but choose "speculative fiction" as the broader frame and include science fiction as a subset of speculative fiction. Speculative fiction is the larger category precisely because it is less defined by boundary-making around the word "science," stretching to encompass related modes such as fantasy and horror, forms of knowledge in excess of white Western science, and more work authored by women and people of color.

In classic science fiction, from the scientific romances and utopian novels of the late nineteenth century through the Hugo Gernsback and John Campbell pulp magazine years, natural disasters dominated stories that imagined disruptions in the weather, though some writers expressed anxieties about scientific experiments going awry and causing catastrophic climate change. Among nineteenth-century climate change novels, Jules Verne's *The Purchase of the North Pole* (1889) is unusual in imagining a geo-engineering scheme hatched by avaricious capitalists in the service of resource extraction. Verne extrapolates from his present, satirizing Americans for trying to capitalize on everything as he imagines them trying to radically alter Earth's climate by changing its axis of rotation to access remote Arctic lands that may hold immense coal reserves, which are "the basis of all our commercial industry." Predicting that before five hundred years are over existing coal reserves will be exhausted, the company justifies making draconian changes to the climate in order to extract this resource. Despite the company's projections, however, a French engineer calculates the force necessary to produce such an effect and predicts it would cause catastrophic disruptions in the earth's crust that would flood most of Asia and other parts of the world. This news causes worldwide panic and efforts to stop the speculative scheme, which is impossible to do because the company has already embarked on the project in an undisclosed location. Luckily, however, the geo-engineer has badly miscalculated, Earth's axis remains unaffected by the firing of the cannon, and in the end Verne reassures the reader that such a man-made change to Earth's axis and climate is impossible because it is "beyond the efforts of humanity" and "our Creator in the system of the universe" will never allow it.

Verne's fiction has enjoyed a long afterlife, frequently reissued in new editions and translations and inspiring many films, from the early days of cinema to the present. One of the earliest ways Verne's work leaped into prominence in popular culture was through the many reprints of his work, including in Hugo Gernsback's pulp magazine *Amazing Stories* (1926–2014), which Gernsback made the home of what he called "scientifiction"—defined as "a charming romance intermingled with scientific fact and prophetic vision." Gernsback was a techno-optimist who fervently believed in scientific progress and justified the existence of a magazine devoted to this "new kind of fiction" by arguing that we "live in a new world" in which "new inventions predicted for us in the scientifiction of today are not at all impossible of realization tomorrow."[32] Verne's work was essential to Gernsback's definition of the genre; he even included a drawing of Verne's gravesite atop each issue's table of contents, claiming access to publication rights from Verne's estate allowed him to disseminate Verne's work to a broad international public. So it was Gernsback who reprinted Verne's "classic" cautionary tale in the September and October issues of *Amazing Stories* during its first year of publication. Ironically, the story's satire of avaricious American capitalists and know-it-all engineers gently undercut the techno-utopianism that *Amazing Stories* usually championed.

Half a century later, British writer J. G. Ballard's four disaster novels of the 1960s, especially *The Burning World* (1964) and *The Drowned World* (1965), which prophetically imagined drought, floods, and other climate changes in most cases caused by industrial pollution and human activity, are usually cited as among the earliest examples of science fictions of climate change. Ballard's work was part of a wave of 1960s books that alerted the reading public to the climate change crisis, including Carson's *Silent Spring*. Meanwhile, most other science fiction novels of the 1960s that imagined a radically changed climate, such as Brian Aldiss's *Hothouse* (1962) and Philip José Farmer's *Flesh* (1960, rev. 1968), returned to the earlier trope of natural disaster in which comets and other natural forces, rather than humans, are responsible for an altered future climate. In the 1960s, Frank Herbert's *Dune* (1965) was an exception and arguably the most ambitious science fiction novel to deal with climate and ecology. Herbert was a journalist who was influenced by Rachel Carson and spoke at the first Earth Day in Philadelphia in 1970. Herbert's intricate and absorbing world-building, set on the harsh desert planet Arrakis, includes an explicit ecological consciousness, "still suits" that turn human body moisture into water, a conflict between imperialist extractors of profit from a scarce resource and locals who try to leave a small footprint and are suspicious of growth as an unexamined ideal, and other elements that inspired the late, great, speculative fiction writer Octavia E. Butler as well as more recent cli-fi authors such as Bacigalupi.

When it comes to people of color's leadership in imagining the future of climate change, Butler's work is a great place to start. She grew up as a working-class Black girl in Pasadena, California, whose mother worked as a maid and by taking in lodgers after her husband died young of a heart attack. Despite her lack of privilege, Butler went on to become a hugely inspiring and formative force in and beyond the world of science fiction. Butler won several major writing awards for novels such as *Kindred* (1979), which hurls its protagonist back to the time of chattel slavery; the *Xenogenesis* novels, repackaged as *Lilith's Brood,* the story of humans surviving a nuclear war by reproducing with aliens and returning to a devastated Earth to make a new world; and her two *Parable* novels, especially *Parable of the Sower,* a "cautionary tale" Butler wrote in the late 1980s and early 1990s. Of the latter, Butler once said that "Global Warming is a character in POS" and while writing the sequel, *Parable of the Talents,* she often reminded herself in research notes to "show the 'Greenhouse World.'"[33] Drawing on the vast collection of material Butler left to the Huntington Library in San Marino, California, next to her hometown of Pasadena, upon her untimely death in 2006, I use Butler's research on the greenhouse effect and global warming and on the disasters of these eras and emerging environmental movements to tell the story of emerging scientific research on climate change in the eighties and nineties, how it was covered in the media Butler collected, and how politicians, the fossil fuel industry, and activists responded to that research. I argue in chapter 2 that this working-class Black woman genius's memory work helpfully illuminates this history even as it models an interdisciplinary engagement with the sciences through Butler's study and research.

Although Butler won major science fiction awards, participated extensively in that world, and is usually classified as a science fiction writer, at times she struggled against the limits of the category because she wasn't sure it completely captured all that she was trying to do. Her writing is so original and ambitious that it often pushes the limits of genre. Even though *Kindred,* to name just one notable example, is a time travel novel, it is also a neo-slave narrative with an abundance of historical texture from Butler's extensive research at the Los Angeles Public Library and in Maryland. Butler called herself a "HistoFuturist," a word of her own devising that means someone who extrapolates from the historical and technological past as well as the present in imagining the future. In notes for a speech she gave about science fiction, she wrote that science fiction "can be one of our methods for looking ahead that I will talk about—not what our future will be, but how we think about it, foresee it."[34] Like Delany, here Butler suggests that science fiction is not really about predicting the future but is rather about the present—how we in the present shape the future that is to come by thinking about it and foreseeing it. In other words, science fiction can help us take hold of the present and think about where things are heading rather than just letting time pass by as our unconscious surround.

But although Butler valued her science fiction community and the genre's usefulness for thinking about and shaping the future, she also saw the utility of the larger umbrella term "speculative fiction." Butler used that term often as well, including in 2004 shortly before her death in a speech she gave at the Black to the Future Science Fiction Festival. Delighted into a "wow" at the existence of a "Black-oriented sf festival," she asked the audience how many of them had "copies of Sheree Thomas's *Dark Matters*" and added that she had especially hoped this crowd would know about Thomas's two "collections of African American speculative fiction of several kinds from as far back as W.E.B. Du Bois." For "this history alone, they're worth having," Butler advised festival participants.[35]

In her groundbreaking 2000 literary anthology, *Dark Matter: Speculative Fiction from the African Diaspora,* editor Sheree Thomas used the term "speculative fiction" to define the genre expansively and to highlight writing that had previously been invisible but was there all along. One striking example is W.E.B. Du Bois's story "The Comet" (1920), which was not considered science fiction or speculative fiction at the time it was published but which is illuminated by situating it in those contexts. The story takes place in Manhattan after a comet passes over the earth and releases noxious gases that kill everyone in the city except a Black workingman and a rich white woman. As in most of the science fiction stories of the time, climate change in "The Comet" is the result of a natural disaster rather than the handiwork of humans, but Du Bois uses this transformative change to think critically about man-made social institutions such as legal segregation: the splitting of the world into black and white halves as a result of the Supreme Court ruling on Plessy v. Ferguson. In this way, Du Bois uses the narrative device of the future to offer a significant distortion of his present and make readers think critically about how the Manhattan of that era was divided into Black and white.

In this book, I am particularly interested in how scholars, writers, artists, and organizers of color have used the terms "speculative fiction" and the "speculative," as well as others such as "futures" and "futurisms," to describe the visionary work they are doing in imagining the future of climate change. "Afrofuturism" became a keyword in Black studies, cultural studies, and American studies after Mark Dery coined it in his 1994 essay "Black to the Future," where he defined it as "speculative fiction that treats African American themes and addresses African American concerns in the context of 20th-century technoculture—and, more generally, African American signification that appropriates images of technology and a prosthetically enhanced future." The term took hold soon thereafter when several influential scholars, artists, and writers started using it to think together about how people of the African diaspora engaged science, technology, and science fiction. Alondra Nelson, a scholar and author of many award-winning books such as *The Social Life of DNA: Race, Reparations, and Reconciliation After the Genome,* started an Afrofuturism listserv in the late 1990s that became a digital hub for the community and the movement. Since then, the word has become increasingly common in popular culture, used to encompass a wide variety of future-facing music, film, literature, and art, as in Ytasha Womack's *Afrofuturism: The World of Black Sci-Fi and Fantasy Culture* (2013).

Other futurisms are also at the heart of this book on imagining the future of climate change. In the introduction to *Walking the Clouds: An Anthology of Indigenous Science Fiction,* editor Grace Dillon, like Sheree Thomas in the case of the writers of the Black diaspora, suggests "Indigenous sf is not so new—just overlooked, although largely accompanied by an emerging movement" (2). Dillon makes comparisons to Afrofuturism as she explains: "Writers of Indigenous futurisms sometimes intentionally experiment with, sometimes intentionally dislodge, sometimes merely accompany, but inevitably change the parameters of sf" (3). One important example for Dillon is Leslie Marmon Silko's novel *Almanac of the Dead* (1991), which she reads as a "near future" story built out of elements of the present, in which "the fight for Indigenous land reclamation and tribal sovereignty is a matter of planetary survival" (217). Dillon "opens up sf to reveal Native presence" (2), making the case for understanding Silko's *Almanac* and other Native texts as Indigenous science fictions and arguing that in Native hands sf has the "capacity to envision Native futures, Indigenous hopes, and dreams recovered by rethinking the past in a new framework" (2). In chapter 1, I build on Dillon's and others' work to read Silko as an important intellectual of climate change, connecting her 1990s "near future" vision to Butler's work as well as to the Indigenous activists who were putting climate justice at the forefront of their struggles during the same period. As we shall see, Silko's vision of movement-building in response to climate change anticipates more recent struggles such as the one over the DAPL, also discussed in chapter 1.

Dillon explores how Indigenous futurisms take a wide variety of cultural forms and often connect to larger movements and worlds. She has done important work herself in making those connections by sponsoring an annual writing contest with a one thousand dollar prize open to "to any emerging writer with an interest in exploring Indigenous issues through the medium of science fiction." Along with others, she also started an "Imagining Indigenous Futurisms" Facebook public group that now has over a thousand members, which provides a space for artists, writers, filmmakers, designers, media makers, activists, and scholars to share insights, exchange information, and highlight work. Indigenous futurisms are at the forefront of efforts to imagine a future of climate change other than that envisioned by the fossil fuel industry and they take many different cultural forms, especially low-cost ones such as videos and photographs captured on cell phones and disseminated across social media such as Facebook and Twitter.

Chicanx and Latinx futurisms also have much to offer in imagining the future of climate change. Thinking about Indigenous people's and people of color's leadership in imagining transnational and international responses to climate change is illuminated by Peruvian American director Alex Rivera's 2008 science fiction film *Sleep Dealer,* which in September 2016 was screened as part of the Climate Change and Climate Justice Film Festival organized by the Institute for the Arts and Humanities at Pennsylvania State University. Widely acclaimed and a favorite on campuses since its release, *Sleep Dealer* is a near future vision of U.S.-Mexico borderlands where transnational capitalists take advantage of an ingenious new technology to drain the labor from workers and redirect it to U.S. work sites while the workers, called "cybraceros," remain in Mexico. The privatization of water in Mexico by a transnational company that builds a pipeline to redirect rural water to the cities is one of the film's important elements. Although Rivera was surprised to be asked to speak on the topic of climate change, his work on imagining a future off the tracks of transnational capitalism resonates with other work on environmental justice, social movements, and imagining the future by scholars in Latinx and American studies. Curtis Marez, for instance, writes about Rivera's film and other work in *Farmworker Futurism: Speculative Technologies of Resistance* (2016), where he analyzes the "limitations," "contradictions," and "critical edge" of specific farmworker visions of the future as well as what he names "moments of materialist futurity, which asks who can expect a future, who cannot, and why" (11). Marez uses the term "speculative" in his subtitle, connecting farmworker speculative futurisms to Afrofuturism, Chicanafuturism, and Jayna Brown's and Alexis Lothian's juxtaposition of "dominant speculation" with "critical forms of 'speculation' that refuse logics of 'profit and power' in order to 'play, to invent, and to engage in the practice of imagining'" (9). And in the 2017 anthology *Altermundos: Latin@ Speculative Literature, Film, and Popular Culture,* editors Cathryn Merla-Watson and Ben Olguín similarly choose the term "speculative" to capture the "creative and resilient ways in which Latin@ cultural producers since at least the 1970s have continued to repurpose and blend genres of sci-fi and fantasy to defamiliarize the ways in which the past continues to haunt the present and future." Instead of focusing only on short stories and novels, these scholars and artists broaden the sphere of the speculative to encompass the arts more broadly as well as the social movements that were energized by them.

Another of my models for thinking about the convergence of climate change, speculative fiction, and Indigenous and people of color futurisms is the collection *Octavia's Brood: Science Fiction Stories from Social Justice Movements* (AK Press, 2015), coedited by Walidah Imarisha and adrienne maree brown, which has gained a wide following of readers both inside and outside the university for its conjoining of science fiction and world-making in writing by participants in movements for social change. The book is inspired by and dedicated to Octavia Butler, out of the editors' "fierce longing to have our writing change everyone and everything we touch." In her introduction, Walidah Imarisha explains that the premise of the collection is that "all organizing is science fiction": that "organizers and activists dedicate their lives to creating and envisioning another world, or many other worlds" and in doing so are "engaging in speculative fiction." She further offers the term "visionary fiction" to "distinguish science fiction that has relevance towards building newer, freer worlds from the mainstream strain of science fiction, which most often reinforces dominant narratives of power" (4). All of the writers in the anthology were inspired by the idea of continuing "Butler's legacy of writing visionary fiction," which Imarisha suggests provides "space" that is "vital for any process of decolonization, because the decolonization of the imagination is the most dangerous and subversive form there is" (4). Our answers about the future of climate change must not come solely from the sphere of science and technology, or they will be too narrow, not capacious enough. The work of the imagination is critical and culture is a crucial contributor to that conversation, not just a handmaiden to the gods of science and technology or a mere reflection of a deeper reality. American studies methodologies, with their emphasis on interdisciplinary thinking and culture's centrality to social movements and the possibility of transformative change, are especially helpful here.[36]

Many of the speculative stories, novels, films, and other futurist cultural forms I center in this study are visionary fictions created by activists and artists who struggle to conceive of worlds that diverge from dominant narratives of power and privilege. They decolonize the imagination by using speculative fiction to break with mainstream stories that center white settlers and fail to imagine deep change. This does not mean that such visionary fictions are optimistic or utopian in a simple way. Often, activists, artists, and writers search for possibilities in the wake of the climate change disaster already upon us rather than turning a blind eye to the many kinds of disaster comprising our current conjuncture's ecological crisis. To understand this, I build on work by science fiction scholars Tom Moylan and Raffaella Baccolini to consider how many critical dystopian texts, especially since the 1990s, offer the glimmer of a utopian horizon as survivors try to create new possibilities in the wake of disaster.[37] I also build on American Studies scholarship that seeks to provide economic and political explanations and contexts for so-called natural disasters such as Hurricane Katrina, which devastated Gulf communities, particularly New Orleans, in 2005. In centering work by Natives and people of color that imagines postdisaster possibilities, I join critics such as Naomi Klein who seek to denaturalize and question the logic of disaster capitalism used by nation-states and corporations to justify the privatization of public services, selling off the environment to the highest bidder, turning places into wastelands, and rendering disposable whole populations.

This book is divided into three chapters that take up three different flashpoints in imagining the future of climate change through visionary speculative fictions and world-making activism. Chapter 1 is entitled "#NoDAPL: Native American and Indigenous Science, Fiction, and Futurisms." The main title refers to the popular 2016 hashtag, #NoDAPL which, in 2016, despite the big-media blackout for most of the year, connected communities and created awareness of the threat posed by the Dakota Access Pipeline to Dakota lands, sacred sites, the Missouri River, and all the creatures who depend on the water. I argue that the skillful mobilization of digital technologies and social media to confront big oil and powerful states are only two examples of robust Indigenous futurisms that encompass many different kinds of high-tech, low-cost cultural productions as well as visionary speculative fictions such as those created by Leslie Marmon Silko and other writers.

The second chapter, "Climate Refugees in the Greenhouse World: Archiving Global Warming with Octavia E. Butler," returns to the 1980s and 1990s, when the problem of global warming first began to be covered widely in newspapers and other media. I argue that through her memory work and archiving activity Butler critically engaged the emerging public climate change conversation and illuminated key blind spots. She did so by centering race, class, and gender and emphasizing the difficult but necessary work of building collectivities in the wake of climate change slow disaster. Butler was ahead of her time in worrying about what she prophetically named "slow disasters," including global warming, which she insisted was not "just an incident like a fire, a flood, or an earthquake" but rather "an ongoing trend—boring, lasting, deadly."[38] Critically commenting on 1980s politicians whom she feared were destroying the planet due to avaricious obtuseness, she warned, "if you notice a slow disaster you have to pay a lot of money, put forward a lot of effort, and wound entrenched interests—who will stop you if they can."[39] Her 1980s and 1990s memory work around climate change in the public sphere resonates with the work of the emerging climate justice movement even as her body of writing and archiving activity raise difficult questions about colonization, community, and coalition-building in imagining the future of climate change through visionary fiction.

The final chapter, "Climate Change as a World Problem: Shaping Change in the Wake of Disaster," begins and ends by focusing on climate justice activist and science fiction writer adrienne maree brown, coeditor of *Octavia's Brood,* who has long used Butler's writing to do powerful work with social movements made up of Indigenous youth, people of color, and white activists. In this last chapter, I emphasize how Indigenous people and people of color have been at the forefront of doing the work of the imagination when it comes to climate justice from the 1990s to the present, in both their movement-building and their speculative and visionary stories, novels, films, web series, and other forms of culture.

ONE

#NoDAPL

Native American and Indigenous Science, Fiction, and Futurisms

In the days leading up to the March for Science on Earth Day, April 22, 2017, more than eleven hundred Native American and Indigenous scientists, scholars, and allies endorsed the "Indigenous Science Statement for the March on Science," authored by four leading Native American scientists and scholars. In this statement, Robin Kimmerer, Rosalyn LaPier, Melissa Nelson, and Kyle Whyte emphasized the concept of Native American and Indigenous science as they encouraged "Indigenous people and allies to participate in the national march in DC or a satellite march." Naming the declaration "Let Our Indigenous Voices Be Heard," the authors insisted on the need to "engage the power of both Indigenous and Western science on behalf of the living Earth."[1] Nelson further elaborated on the concept of Indigenous sciences in an interview: "To successfully address our world's pressing ecological issues, it is critical that we look to the multiple place-based and time-tested sciences of Indigenous peoples."[2] The use of the term Indigenous science, like the multiple Indigenous science organizations—including the American Indian Science and Engineering Society, the National Coalition of Native American Language Schools and Programs, and the Society Advancing Chicanos/Hispanics and Native Americans in Science—that endorsed the March for Science, is itself a significant theoretical claim. The idea that Indigenous peoples practice sciences and have deep historical knowledge that is often "place-based" is an important intervention in a settler society predisposed to discount Indigenous perspectives. The critique of versions of Western science based on narrow, linear notions of progress and development inseparable from histories of colonialism and racial hierarchies is also noteworthy. The declaration goes on to argue that both Indigenous and Western sciences, working together for the sustainability of the earth, are necessary at the current conjuncture.

In the opening, the authors emphasize that although "Western Science is a powerful approach, it is not the only one." Calling the Earth Day event a "march not just for Science but for Sciences," they remember that "long before Western science came to these shores, there were Indigenous scientists here": "Native astronomers, agronomists, geneticists, ecologists, engineers, botanists, zoologists, watershed hydrologists, pharmacologists, physicians and more—all engaged in the creation and application of knowledge which promoted the flourishing of both human societies and the beings with whom we share the planet." This history of Indigenous science is relevant to our present, they insist, because it "supported indigenous culture, governance and decision making for a sustainable future—the same needs which bring us together today." It also offers "a wealth of knowledge and a powerful alternative paradigm by which we understand the natural world and our relation to it," providing "key insights and philosophical frameworks for problem solving that includes human values." The latter are indispensable for facing "challenges such as climate change, sustainable resource management, health disparities and the need for healing the ecological damage we have done." Their demands include "greater recognition and support for tribal consultation and participation in the co-management, protection, and restoration of our ancestral lands" as well as "enhanced support for inclusion of Indigenous science in mainstream education, for the benefit of all." In these ways, the authors and signers of this document "envision a productive symbiosis between Indigenous and Western knowledges that serve[s] our shared goals of sustainability for land and culture" while emphasizing that "this symbiosis requires mutual respect for the intellectual sovereignty of both Indigenous and Western sciences."

In this chapter, I build on scholarship in Native American and Indigenous studies and American studies, cultural forms produced by social movements disseminated through the Internet, and journalism and other media to further elaborate on the possibilities for such a "productive symbiosis" as well as the concepts of Indigenous science, fiction, and futurisms that are crucial for confronting the imminent disaster of climate change today. I also suggest that Indigenous science, fiction, and futurisms have converged to shape struggles over the DAPL as well as other struggles over water, oil, and resource extraction throughout the world.

From late spring through fall 2016, while the candidates for U.S. president failed to address climate change, a series of major events in the history of imagining the future of climate change was taking place. On April 1, tribal citizens of the Standing Rock Lakota Nation and other Lakota, Nakota, and Dakota citizens founded a spirit camp along the proposed route of the 1,172-mile DAPL. They objected not only to the pipeline but also to the use of the word "Dakota" to name it and the company that hoped to transport fracked oil from the Bakken oil fields across three states to refineries in Illinois. Dakota Access, LLC is a subsidiary of the Dallas-based company Energy Transfer Partners, which owns and operates more than 62,500 miles of natural gas and liquids pipelines. Fracked oil is created by hydraulic fracturing of tar sands and is more volatile and damaging to local ecosystems than conventional oil extraction. There are huge gaps in our knowledge of how spilled tar sands oil behaves in water and fracked oil may be more corrosive to pipeline systems than oil. Naomi Klein reports that "a growing body of independent, peer-reviewed studies is building the case that fracking puts drinking water, including aquifers, at risk."[3] Evidence also suggests that fracking causes small earthquakes. There are significant reasons, then, to worry about Energy Transfer Partners' pipeline, which passes under the Missouri River. This corporation has powerful friends, however: Trump has significant stock holdings invested in the pipeline and CEO Kelcy Warren donated hundreds of thousands to Trump, the Trump Victory Fund, and the Republican National Committee in 2016.

To provide the material basis for resisting Energy Transfer Partners' pipeline, LaDonna Brave Bull Allard, the tribe's historic preservation officer, cofounded the Sacred Stone Camp on her land in April 2016. The camp was called Iŋyaŋ Wakháŋagapi Othí, translated as Sacred Rock, "which was the pre-colonial name of the Cannonball area," and became the site of emergence for "a historic grassroots resistance movement" that was "determined to stop the pipeline through prayer and nonviolent direct action."[4] When Allard heard construction would start on the pipeline, which would be routed near her water well and her son's grave, she posted a video message on Facebook asking for help. That post went viral and soon so many people showed up that an overflow camp had to be established across the river.

Earlier, in March 2016, officials at the Environmental Protection Agency and two other federal agencies raised serious environmental and safety objections to the North Dakota section of the pipeline, warning "crossings of the Missouri River have the potential to affect the primary source of drinking water for much of North Dakota, South Dakota, and Tribal nations." But the Army Corps of Engineers dismissed these concerns and relied on an environmental assessment prepared by the company itself when it issued a fast-track permit July 25 for pipeline construction to continue.[5] Finally, after three easements for water crossings on Lake Oahe, Lake Sakakawea, and the Mississippi River were granted by the Corps, the Standing Rock Sioux filed an emergency injunction request to stop construction. Two weeks later Energy Transfer Partners countersued Standing Rock chairman Dave Archambault II and others for blocking construction. In the following weeks, thousands of people, with more than three hundred tribes represented, came to join forces with Standing Rock water protectors.

This fight for Indigenous land reclamation, sovereignty, and survival is partly rooted in longer histories and other times that coexist with and shape the present. Allard calls attention to the significance of earlier struggles over settler colonialism and resource extraction on Native lands for the current conflict. Emphasizing that she "can't forget" the 1863 "Whitestone Massacre" that took place almost 150 years earlier, Allard recalled in a magazine piece how her great-great grandmother Nape Hote Win (Mary Big Moccasin) survived the "bloodiest conflict between the Sioux Nations and the U.S. Army ever on North Dakota soil," in which "an estimated 300 to 400 of our people" were killed, "far more than at Wounded Knee."[6] As part of a "broader U.S. military expedition to promote white settlement in the eastern Dakotas and promote access to the Montana gold fields via the Missouri River," U.S. troops in Dakota territory attacked a peaceful camp with large numbers of women and children after a buffalo hunt and killed hundreds of people. Nine-year-old Mary, Allard's ancestor, was shot in the hip, found by a U.S. soldier the next morning, and taken to a prisoner-of-war camp, where she remained until 1870.

For Allard this story of survival in the face of U.S. military violence in the service of white settlement and resource extraction is entangled with the Standing Rock struggle, as well as the 1950s moment when the U.S. Army Corps of Engineers "dredged the mouth of the Cannonball River and flooded the area" as "they finished the Oahe dam." At that time, she recalls, "they killed a portion of our sacred river," desecrated "burial sites and Sundance grounds," and made the river water unsafe to drink. Noting that of the 380 archaeological sites that face desecration along the pipeline route, 26 are situated at the confluence of the Missouri and Cannonball Rivers, Allard charged that history was repeating itself: "Again, it is the U.S. Army Corps that is allowing these sites to be destroyed."

But if these sites are not protected, Allard warns, "our world will end," because that world is inseparable from the particularities of place. When she looked at the pipeline map, she knew her "entire world was in danger." Because "we are the river, and the river is us," the Oceti Sakowin have "no choice but to stand up," together with other tribes gathered at Standing Rock, and "demand a future for our people." In Fall 2016, as the U.S. election season heated up, the Standing Rock Sioux and their allies projected a different future of climate change by trying to protect the water, the land, and the whole web of life that depends on the river, partly through their adept use of social media. Suddenly the Lakota saying *Mni Wiconi* or "Water Is Life" began to spread across social media platforms such as Facebook, Twitter, and Instagram despite the big-media blackout. The water protectors were skilled at using high-tech cultural forms to organize in the present and imagine a future connected to the past beyond the global fossil fuel economy. Indigenous women such as Allard were leaders of this effort, speaking at the media tent at the camp and using social media to organize networks that spread news of the pipeline struggle far and wide.[7]

Standing Rock youth and their skilled use of social media were also crucial. In July, Bobbi Jean Three Legs, Montgomery Brown, and Joseph White Eyes, all in their twenties, organized a nearly two thousand mile, intertribal relay run from North Dakota to Washington, D.C. to deliver a petition against the pipeline with over 160,000 signatures to the White House and the Army Corps of Engineers.[8] They also launched a massive media campaign, "Rezpect Our Water," that includes a website with an anti-DAPL petition, videos made by youth, and letters to the Army Corps of Engineers. The Standing Rock Youth Council, made up of "Youth from all nations, tribes, and races," emerged at this time to "protect land, water, and treaty rights," "work towards the end of environmental racism," take "Non Violent Direct Action," and advance their "voices in decisions made about the future of Indian Country." They situated themselves "in the tradition of our elders and the American Indian Movement in coming together nationally and internationally to form a solidarity movement that builds people power," often by effectively using hashtags and making connections on social media platforms.[9]

The efforts of the water protectors and their allies to collectively imagine a different future were also strikingly evident in the camp itself as well as at other camps such as Oceti Sakowin Camp, which was set up across the Cannonball River to accommodate huge numbers of people who arrived in August. Both Sacred Stone Camp and Oceti Sakowin included schools for kids. Standing Rock teacher Alayna Eagle Shield started the school at Oceti Sakowin and was soon able to draw on many talented Indigenous teachers who came to Standing Rock. Using donated tablets with solar chargers, the schoolchildren, who said they were tired of how reporters told their stories, created their own films, involving research and interviews, while taking classes in math, science, and Lakota values and language.[10] Education was also available for older people: nonviolent civil disobedience and "direct action" principles were proclaimed on signs and taught to those who wanted to learn them.

The Industrial Workers of the World used the phrase "direct action" in 1910 to describe actions taken by workers themselves, including strikes and demonstrations, as opposed to actions taken by politicians and other mediators representing workers by negotiating or making laws. The phrase has taken on added meanings as it has traveled across space and time, playing a significant role in the U.S. civil rights, antiwar, and environmental movements.[11] At a moment when political, economic, and social change from above has become increasingly hard to imagine, direct action has moved to center stage in the world-making projects of many contemporary social movements.

The ways in which the camps of the water protectors and their allies made another world through direct action were not limited to schools and education, however, although both were important. Camp kitchens produced three meals a day, often made up of Indigenous foods, for hundreds of people.[12] The Standing Rock Medic and Healer Council, despite limited resources, provided health care to those assembled and medical professionals from as far away as Cuba came to express their "solidarity with the Sacred Water Protectors on the front lines of the current human rights and ecological crisis occurring right now in North Dakota."[13] The water protectors not only made a world that provided schooling, food, medical care, and other necessities of life but the world came to Standing Rock. On August 22, 2016, after protestors used direct action to block a construction site at Cannon Ball, North Dakota, members of more than 280 tribes headed there to provide support. By September 11, the *New York Times* was calling it "the largest, most diverse tribal action in at least a century, perhaps since Little Bighorn."[14] Franco Viteri, former president of the Confederation of Indigenous Nationalities of the Ecuadorian Amazon (CONFENAI), arrived with two other members of his Kichwa community. All had participated in decades-long struggles against multinational oil companies and the Ecuadorian government's efforts to drill in the Amazon. They told reporters they saw many other Indigenous people from Latin America at the camp and recalled speaking with people from Honduras, Peru, and El Salvador.[15]

In response to these events, the American Studies Association's 2016 Conference in Denver featured a special Saturday night session entitled "#NoDAPL: Indigenous Dispossession on the Missouri." The panel, which was chaired by Robert Warrior, the distinguished scholar of Native and Indigenous literature who was president of the American Studies Association that year, focused "on the recent protests along the Missouri River initiated by the Lakota people of the Standing Rock reservation" and emphasized Indigenous people's crafting of futurisms through social media and other technologies by using the hashtag #NoDAPL as the main title. All of the speakers connected the #NoDAPL struggle to longer histories of settler colonialism and Indigenous survivance and world-making, including Marcella Gilbert, a Cheyenne River Sioux community organizer, who emphasized the importance of Native women's imaginings of the future in the long history leading up to the present. Gilbert testified to how she grew up in the American Indian Movement (AIM) and had relatives involved with the 1979–81 occupation of Alcatraz, when Red Power activists claimed the island under the terms of the 1868 Fort Laramie Treaty, which gave Native people rights to unused federal property on Indian lands. Everybody went out there to live, she remembered, like at Standing Rock. She asked the audience to remember how in October 1972, AIM and other Native groups had organized a protest in Washington, D.C. known as the "Trail of Broken Treaties" that was one of many in that period, including two takeovers of Mount Rushmore and the 1973 occupation of Wounded Knee on the Pine Ridge Indian Reservation. She reminded everyone of other antecedents of the present, including "survival schools" created in Minneapolis and St. Paul by AIM members in the early 1970s and the international work of the 1970s and 1980s, much of which focused on sustainable living and different ways of surviving off the grid. She recalled other AIM world-making projects such as KILI Radio, created in 1983 as the very first Indian-controlled, Indian-owned, and Indian-run radio station in the United States as well as medical clinics and tribal colleges, created on the backs of movements and women who knew how to get things done. These times have shaped both the present and the future, and Native American and Indigenous studies scholars, artist, writers, and other knowledge producers have much to teach us about long histories of social movement activity and future-facing connections across space and time.

People from Indigenous nations throughout the Americas have joined climate activists, members of the Black Lives Matter movement, and other allies to insist that construction be stopped on the Dakota Access and Keystone XL pipelines. As well, coalitions have arisen in response to the extraction of oil from the Alberta tar sands in Canada; to the U.S. oil pipelines that big capitalists are trying to run through what they imagine as flyover places, or places where disposable people live who don't really matter; and to the struggles over deforestation, water, and oil that have long been going on and continue today in Ecuador, Peru, and other parts of the Americas. Traci Voyles calls the transformations such places undergo "wastelanding," defined as the extraction of resources in racialized spaces that combined with environmental racism renders "space marginal, worthless, and pollutable."[16] Indigenous waterways have suffered many of the worst environmental catastrophes. Responding to such projects of resource extraction as well as the siting of waste dumps and hazardous polluting industries in close proximity to communities of Indigenous people and people of color, the water protectors and their allies thus contribute to a revitalized politics of place at a time when some have argued the conditions of postmodern life make place-based organizing unsustainable. At the same time, they reveal how place-based struggles can connect people who are widely separated geographically but bound together in confronting common antagonists and sharing common goals.

A good deal of the emerging scientific research suggests that Indigenous people and people of color will be among those most affected by climate change impacts. Indigenous people in the United States have also been among those most impacted by the environmental consequences of extractive industries. According to Kyle Whyte, because environmental consequences have been "part of the experience of every single tribe" in the United States, Indigenous people "have been at the forefront of taking action against extractive industries."[17] Whyte lists numerous examples leading up to the struggle over the DAPL, especially the use of direct action together with legal and political approaches to stop the Crandon mine in Wisconsin, which would have affected the "lands and waters of the Mole Lake Sokaogon Tribe, the Forest County Potawatomi Tribe, and other nearby tribes, such as the Menominee." In order for Indigenous people to survive, Whyte concludes, many communities are deciding they have to say no to the fossil fuel industry, especially when Indigenous communities in the Arctic, the Pacific, and the Gulf of Mexico are "experiencing climate change impacts right now that pose threats of the highest severity." Although the severity of this threat cannot be overstated, Whyte also emphasizes the significance of the emergence of "a global indigenous people's movement, a lot of which focuses on environmental protection," noting that "Indigenous people are the largest group of non-state actors to have a presence in the United Nations, with three different institutions devoted to indigenous rights."

INDIGENOUS EPISTEMOLOGIES, INTERSECTIONALITY, AND RESISTANCE TO SETTLER COLONIALISM

There is an important and irreducible particularity to the struggle of the Standing Rock Sioux at this site. As scholar-activist Edward Valandra explains, the water protectors push beyond environmentalist paradigms in a number of ways, including through understanding the river and the water as living persons.

Although the Western way of life "denies" and "defies" the idea of water having personhood, he points out that the U.S. government "arbitrarily recognizes fictional entities like corporations as real persons" while historically denying personhood to real people such as slaves. Valandra argues that "our kinship obligates us to protect water from egregious harm."[18] While climate change is certainly important, the Indigenous struggle also involves long histories of settler colonialism and conflicts over land, water, oil, and other resources that had great costs for Native people while creating profits for settlers. Many Indigenous people throughout the world are connected through experiencing similar struggles over land, water, oil, and pipelines against transnational corporations and states that wish to extract value from them while wastelanding their communities.

These struggles also intersect with those of participants in other important social movements today. Critical race theorist and Black feminist Kimberlé Crenshaw coined the term intersectionality in 1989 to call attention to overlapping vectors of identity and how race and gender in particular interact, though class, sexuality, citizenship status, and other elements of identity are also salient. In September 2016, Black Lives Matter (BLM) issued a statement of solidarity with the water protectors, calling the Standing Rock movement "a movement for all of us," based on the principle that "water is life," "led by warriors, women, elders, and youth."[19] Declaring that the "water protectors who are protesting the DAPL are engaged in a critical fight against big oil for our collective human right to access water," the BLM network advised members of the movement that "this is not a fight that is specific only to Native peoples—this is a fight for all of us and we must stand with our family at Standing Rock." They explained that this was partly due to the intersectionality of identity and collective struggle: "As there are many diverse manifestations of Blackness, and Black people are also displaced Indigenous peoples, we are clear that there is no Black liberation without Indigenous sovereignty." But BLM's statement also emphasized how solidarities emerged from connected histories of environmental racism, recognizing that "Environmental racism is not limited to pipelines on Indigenous land, because we know that the chemicals used for fracking and the materials used to build pipelines are also used in water containment and sanitation plants in Black communities like Flint, Michigan. The same companies that build pipelines are the same companies that build factories that emit carcinogenic chemicals into Black communities, leading to some of the highest rates of cancer, hysterectomies, miscarriages, and asthma in the country." Many other progressive organizations also made connections between their work and the confrontation with the fossil fuel industry and the state at Standing Rock, comprehending the threat to water as part of the exploitation of marginalized and working-class people by the powerful.

Although the Standing Rock struggle created other worlds and futures and connected movements with intersecting struggles, however, it was not a utopia. Utopias depend upon a physical or spatial separation from dominant and mainstream communities that allows another way of living to flourish, but this alternate world was situated within the settler colonial space of the United States and thus subject to force and violence from the Morton County Sheriff's Department and other police as well as DAPL security personnel. The space was also layered with histories of settler colonialism and struggles over the dispossession of Indigenous people. These long temporal roots of the Standing Rock struggle in histories of nineteenth-century empire-building and settler colonialism are usefully emphasized by the Standing Rock Syllabus Project, published online on behalf of the New York City Stands with Standing Rock Collective by Matt Chrisler, Jaskiran Dhillon, and Audra Simpson in September 2016 as part of a New School for Social Research Public Seminar. The syllabus includes a timeline of U.S. settler colonialism that extends backwards to Columbus's expeditions in the Antilles/Caribbean in 1492. In their statement and in the linked materials, the Collective also emphasizes the nineteenth-century history of contact between Europeans and the Oceti Sakowin, extending from the Louis and Clark expedition in 1803 through the Fort Laramie Treaties, signed in 1851 and 1868 by the United States, the Sioux, and other tribes, which would be repeatedly violated by the United States; the Great Dakota Uprising of 1862; the 1876–77 Great Sioux War; the breakup of the Great Sioux Reservation in 1889; the Ghost Dance that emerged in response; and the 1890 Battle of Wounded Knee, where the U.S. Army massacred 250 to 300 Lakota, mostly women and children. As well, the syllabus highlights several relevant twentieth-century flashpoints, notably the 1944 New Deal passage of the Pick-Sloan Missouri Basin Plan, the massive water infrastructure project along the Missouri River and its tributaries that entailed the Army Corps of Engineers violating once more the Fort Laramie Treaties as well as the "Winters Doctrine supporting the sovereignty of tribal lands, consultation, and access to water."[20] Just a few years later, in 1948, construction was begun on Lake Oahe Dam, which was completed in 1962 and destroyed more Native land than any other water project in the United States, eliminating 90 percent of timber land on the Standing Rock Sioux and Cheyenne Sioux reservation lands, along with grazing and agricultural land. The Collective's syllabus recognizes that "Indigenous peoples around the world have been on the frontlines of conflicts like Standing Rock for centuries" and thus longer global histories of empire and colonialism are necessary to understand the current conjuncture.

The water protectors put their bodies on the line in ways that resonated with earlier Indigenous struggles against the settler state and its extractive economy as well as mid-twentieth-century civil rights struggles and Indian Wars of the nineteenth century. On September 3, 2016, after tribal officials said that construction crews had destroyed sacred burial and cultural sites on private land in southern North Dakota, protestors confronted the crews and were attacked with dogs and pepper spray, recalling how dogs were used to attack civil rights protestors in Birmingham, Alabama in the 1960s as well as in slavery contexts throughout the Americas in earlier centuries.[21] On October 27, police from seven states in riot gear with Humvees emitting high-pitched "sound cannon" blasts and helicopter support, pepper sprayed and arrested 141 people for setting up an "illegal roadblock" with wood and bales of hay and for trespassing by camping on pipeline property, locked them in cages in the garage of the Morton County Correctional Center, and wrote identification numbers on their arms.[22] Many of those arrested used Facebook and other social media to circulate photographs, videos, and testimony about the militarized police action, which included entering the tipis that water protectors had made in the pipeline's path. "It looked like a scene from the 1800s, with the cavalry coming up to the doors of the tipis, and flipping open the canvas doors with automatic weapons," one of them said.[23] Then on November 20, an army of riot police used munitions, chemical agents, a water cannon, and hoses to blast water protectors on the reservation in subfreezing temperatures, causing hypothermia and severe injuries that were, once again, documented by Facebook videos in scenes "that looked like a siege." The use of fire hoses, which reminded viewers of infamous images of police training high-powered fire hoses on schoolchildren who were marching for equality in Birmingham in the 1950s, had been phased out fifty years ago.[24] Along with attacks by dogs and small armies of police, soldiers, and private security workers and assaults with high-pressure hoses spraying water and chemicals, the drones, helicopters, and planes that patrolled the sky were another everyday reminder of the militarized settler colonial order that constrained and punished the water protectors' efforts to make a different world. In another remarkable neocolonial convergence, the Army Corps of Engineers announced they would close the camp on George Custer's birthday, December 5, but backed off after a public outcry.

In the aftermath of the assault on the water protectors and threats from the North Dakota governor to evacuate the camp, thousands of veterans came to Standing Rock to help defend them.[25] On the day before the December 5 evacuation deadline, however, the Corps abruptly changed course, announcing that the easement to cross under Lake Oahe was denied and that they would explore alternate routes by producing an Environmental Impact Statement with full public input and analysis. As the water protectors celebrated this dramatic victory, many veterans, including co-organizer Wes Clark Jr., turned attention to the water crisis in Flint, Michigan, making connections between the two. Congresswomen and veteran Tulsi Gabbard from Hawai'i came to Standing Rock and told National Public Radio that what "the people here" are fighting "is not different from the unfortunate lead poisoning that the people of Flint, Michigan, have been enduring, and is not different from a threat that we face in Hawai'i right now where one of our largest water aquifers sits beneath a fuel storage tank carrying millions and millions of gallons of fuel that's actually leaked already tens of thousands of gallons of fuel, threatening the water quality in our aquifer."[26] In these ways, the Lakota saying "Water Is Life" and the world created by the water protectors connect struggles over the future of climate change throughout the nation and around the world.

NATIVE SLIPSTREAM AND TIME-BENDING

In insisting on the significance of long histories and connections among different flashpoints in time, the New York City Stands with Standing Rock Collective organized its syllabus in ways that resonate with Native slipstream, which Grace Dillon, editor of *Walking the Clouds: An Anthology of Indigenous Science Fiction,* defines as "a species of speculative fiction within the sf realm" that "infuses stories with time travel, alternate realities and multiverses, and alternative histories." Native slipstream "views time as pasts, presents, and futures that flow together like currents in a navigable stream," thereby "replicating nonlinear thinking about space-time."[27] Dillon notes that Anishinaabe writer and scholar Gerald Vizenor used the term "slipstream" in 1978, eleven years before cyberpunk writer Bruce Sterling, who is usually credited with coining it in a column in the fanzine *Eye.* She observes that its roots go deeper, emerging in the early 1900s in aeronautics industry lingo and in 1983 in the name of the literary publishing house Slipstream Press. Dillon includes in *Walking the Clouds* Vizenor's 1979 story "Custer on the Slipstream," situating it as "a seminal work that brings together two well-known yet never before aligned staples of Native American resistance and science fiction legerdemain: Custer as the poster child of the limitations of white oppression and time travel through alternate realities" (17).

Written just a year earlier, in 1978, Vizenor's first novel, *Darkness in Saint Louis: Bearheart* (1978), mixes different times and flashpoints in Native history, fitting the definition of speculative fiction both in its slipstream qualities as well as its form as a story of the near future. Extrapolating from the 1970s oil crisis, Vizenor imagined a world built out of elements of his present, in which "national supplies of crude oil" dwindle "to nothing."[28] The United States has become a postapocalyptic wasteland because the state was "incapable of negotiating trades or developing alternative fuels." In 1990, Vizenor revised the novel and published it under the title *Bearheart: The Heirship Chronicles.* No doubt he found as timely as ever this story of the U.S. government exploiting "native lands and natural resources" with renewed intensity after "economic power had become the religion of the nation." In it, the state takes half of the nation's standing timber, mostly located on Native lands, even as it opportunistically uses disaster to authorize violent, antidemocratic "restrictions and regulations on travel and the use of fuels." Early on, a radio voice announces that all citizens will be issued "bionic residential identification cards" and that nobody will be "permitted to travel without government authorization." Citizens found cutting or selling federal timber will be executed. Although Vizenor does not imagine climate change, his speculative fiction of a near future where the U.S. government invades reservations to extract resources in the dying days of the fossil fuel economy responds to a long history of such violent resource extraction even as it anticipates the world we are living in now, thereby mixing up different times to yield insights in ways that are characteristic of slipstream as well as Native epistemologies.

One year after Vizenor published the revised *Bearheart,* Laguna writer Leslie Marmon Silko in her great 1991 novel *Almanac of the Dead* offered a powerful speculative fiction of climate change and transnational alliances from below that prophetically illuminates recent struggles over oil and water throughout the hemisphere, including the DAPL. It also anticipates recent cli-fi novels such as Bacigalupi's *The Water Knife* (2015), which makes dystopia the motor of its near-future action in the post-climate change U.S. Southwest. Although *The Water Knife* also centrally focuses on struggles among Native people, states, and corporations over water rights, however, Silko's near-future vision is more complexly shaped by Indigenous people's organizing and leadership in emerging environmental movements.

Moving back and forth between moments in time that flow together is important at the end of *Almanac,* when Silko's character Sterling returns home to Laguna Pueblo in New Mexico after spending many years away in Tucson, Arizona. Sterling wanders out to an open-pit uranium mine and reflects on how he "he had not understood before why the old people had cried when the U.S. government had opened the mine" decades ago. He is newly struck by the mine's monstrous growth over time, how it "had devoured entire mesas." He recalls how the old people who cried soon thereafter saw "the first atomic explosion," followed "weeks later by the bombs that had burned up half a million Japanese." No wonder Sterling suddenly time travels—his temporal perspective abruptly shifts—as he imagines the land a thousand years ago, "when the rain clouds had been plentiful." Or that Sterling also enjoys looking ahead, "five hundred years or so," to the fulfillment of Lakota prophecy predicting that buffalo will reappear after all of the groundwater has been sucked out of the Ogallala Aquifer and only people who know "how to survive on the annual rainfall" remain.

Another significant plot in *Almanac* involves Mafia wife and Arizona, real estate developer Leah Blue, who imagines Venice, Arizona as a dream city of the twenty-first century, complete with a maze of canals. She puts her faith in geo-engineering schemes to realize her vision, unconcerned about whether all the water is being used up "because science will solve the water problem of the west. New technology. They'll have to." Meanwhile, she plans to drill deep wells on lands whose water rights are claimed by Indigenous people in Nevada. Although Natives file a lawsuit, Leah persuades her Mafioso husband to convince a judge to dismiss it and thereby acquires the permits for herself: "Indian tribes or ecologists might try to sue to stop her deep wells later, but by then the deep wells would be flowing in Venice, Arizona." In this way, Silko's near-future vision is shaped by elements of her present that became even more prominent in the future.

Dillon observes that Silko's novel "follows the early-eighties trend in cyberpunk to set sf tales in the near future rather than millennia removed," projecting an imminent future where "Indigenous Peoples all over the Americas are converging: they come from the lake country of Canada, from the alleys of New Orleans and the swamps of Florida, from the islands of the Caribbean, representing a diversity of Native tribes and bands, Louisiana's Black Indians, and the descendants of rebel slaves of African and Native blood who together cross racial and national boundaries." They also come together at an international conference "called by natural and Indigenous healers to discuss the earth's crisis," which includes disappearing rain clouds, "terrible winds and freezing, and burning, dry summers." In the novel's climax, "eco-warriors" partner with a Hopi leader to destroy Glen Canyon Dam and liberate the Colorado River to take back the earth from the "new enemies," the "space station and biosphere tycoons" who are "depleting rare species of plants, birds, and animals so the richest people on earth" can "bail out of the pollution and revolutions and retreat to orbiting paradise islands of glass and steel." In these ways, Silko projects a near future in which Native people and their allies join forces across national borders and boundary-lines to foil the geo-engineering schemes and escape plans of the rich despoilers of the earth. As Dillon puts it, Silko "thereby rewrites the history of the Americas by predicting an Indigenous revolution that straddles the U.S. and Mexican border" and insists that "the fight for Indigenous land reclamation and tribal sovereignty is a matter of planetary survival" (216–17).

Silko's vision of people of color coming together in response to Earth's crisis was partly realized in 1991, the year the novel was published, by the First National People of Color Environmental Leadership Summit in Washington, D.C., a major marker of people of color's significant role in imagining the future of climate change. More than three hundred African, Latino, Native, and Asian Americans from all fifty states attended the summit, which also included delegates from Puerto Rico, Canada, Central and South America, and the Marshall Islands. Extrapolating from that moment, Silko depicts an historic near-future meeting of Native peoples and people of color from all over the world to plan a revolution that demands direct action in order to save the planet. Silko thereby imagines a transnational alliance among Native people and their allies that presages the one at Standing Rock and others throughout the hemisphere since the 1990s.

Indigenous people and people of color provided important climate change leadership long before the 1990s, however. And throughout that decade they continued to shape emerging international debates about it in ways that resonate with Silko's speculative fiction. One particularly significant example is that of Hopi leader Thomas Banyacya (1909–99), whose history connects him both to Sterling as well as to Silko's Barefoot Hopi, who is allied with ecological activists and a main mover of the revolutionary conference near the novel's end.[29] In the story that Banyacya repeated for the rest of his life, in 1948, after the U.S. used nuclear weapons against Japan, Hopi elders appointed four young men who spoke English as messengers to bring the Hopi prophecies to the attention of the English-speaking world. Banyacya had just finished serving a seven-year prison sentence for refusing to register for the draft during World War II and remembers that he resisted taking on this new duty but finally accepted it.

Banyacya sometimes made the news in the 1970s, when Hopis joined forces with environmentalists in trying to stop Peabody Coal Corporation from strip-mining Black Mesa. Banyacya spoke out against the massive relocation of Navajo people from Hopi land because of the human costs of this move, which were severe, and out of concern that mining companies would acquire access to coal, uranium, and oil shale on these lands. In spring 1972, he traveled to the first UN Conference on the Human Environment in Stockholm, hoping to bring the strip-mining of Black Mesa to the world's attention. Stewart Brand of the Whole Earth Catalog had offered to sponsor fifteen Native Americans to come to the conference and present their perspectives on Mother Earth. According to historian, artist, and activist Jack Loeffler, at this conference Hopis and Navajos met with other Indigenous leaders from around the world, thereby establishing "a network that continues today."[30] In the following years, Banyacya traveled to Japan, Russia, Central Europe, and elsewhere to communicate the Hopi prophecy while declining to apply for a U.S. passport. Instead he created one identifying him as a citizen of the Hopi nation.

For many years, Banyacya tried to bring the Hopi message to the United Nations but was not given permission to speak. Finally in December 1992, he was invited to address the General Assembly when it was in recess. Hardly anyone was present and you get the idea they were just humoring him. He began by speaking in Hopi, showing his awareness he was being humored and snubbed by observing "only a handful of United Nations Delegates are present to hear the Motee Sinom [Hopi for First People] from around the world who spoke here today."[31] He then switched to English, reminding his small audience that "the United Nations stands on our native homeland." He proceeded to charge the UN with ignoring Indigenous voices and advised the assembly to get to work trying to reverse the damage humans had done to the earth: "The United Nations talks about human rights, equality, and justice and yet the Native people have never had a real opportunity to speak to this assembly since its establishment until today." He also spoke of "loud warnings" given by "increasing floods, more damaging hurricanes, hail storms, climate changes and earthquakes as our prophecies said would come." He told them that the elders had requested that "during this International Year for the World's Indigenous Peoples, the United Nations keep that door open for spiritual leaders" to "come to speak to you for more than a few minutes as soon as possible." In another 1992 speech at the World Uranium Hearing in Salzburg, he specifically pointed to fossil fuel extraction as a world problem that the UN must address in order to meet its greater goal of preserving world peace.

In *Almanac,* a Barefoot Hopi with "radical schemes" is a prime mover of the "international convention" called by "natural and indigenous healers to discuss the earth's crisis" at the end of the novel. A main concern is the changing weather. At the convention, the Barefoot Hopi joins Wilson Weasel Tail, a Lakota "raised on a small, poor ranch forty miles from the Wounded Knee Massacre site," who speaks eloquently, warning of the earth's "outrage": "the rain clouds no longer gather; the sun burns the earth until the plants and animals disappear and die." The Hopi also delivers the conference keynote, in which he speaks of the "international coordinated effort" in which he plays a central part, traveling to Africa, Asia, and "around the world to meet with indigenous tribal people." He succeeds in creating a transnational coalition by emphasizing "similarities between the tribal people of Africa and the tribal people of the Americas," especially in terms of settler colonial land loss to whites. The Barefoot Hopi also "has a growing number of disciples inside and outside jails and prisons" and speculates about a "national or even a multinational prison uprising" coordinated with the activities of eco-warriors. Announcing that the eco-warriors have liberated the Colorado River, he wards off charges of terrorism by turning the charge around and accusing the United States of terrorism, defined as poisoning the water and air. He predicts earthquakes, tidal waves, landslides, drought, and wildfire before it's all over, but also promises that "a force [is] gathering," coming from the South, to "counter the destruction of earth."

Vizenor's and Silko's speculative fiction of the 1980s and early 1990s offers powerful Indigenous futurisms in the face of the battles that were taking place over resource extraction, environmental racism, settler colonialism, and climate change. At the same time that neoliberals were deepening imminent climate change disaster, not only was Indigenous speculative fiction challenging the Western progress narrative, but movements led by Indigenous people and people of color were already leading the way in imagining a different future.

THE ENVIRONMENTAL JUSTICE MOVEMENT AND THE INDIGENOUS ENVIRONMENTAL NETWORK

The year 1990 was especially significant for Indigenous people's future-facing social movement formation. In that year, the Southwest Organizing Project, a network of activists and organizations, sent letters to the Group of Ten—the most prominent mainstream environmental organizations, including the Sierra Club, the National Audubon Society, and the National Wildlife Confederation—asking them to address the exclusion of communities of color from the environmental movement. Writing in hopes that "through dialogue and mutual strategizing we can create a global environmental movement that protects us all," the authors charged that "there is a clear lack of accountability by the Group of Ten environmental organizations towards Third World communities in the Southwest, in the United States as a whole, and internationally."[32] When the letters did not produce an adequate response, organizers called for a summit. In early April, they held the People of Color Regional Activist Dialogue for Environmental Justice, a gathering of Third World political activists from eight Southwestern states, including Native nations. Their purpose was to begin analyzing the realities of environmental injustice facing Third World communities in the Southwest and to develop regional strategies. At this meeting, the Southwest Network for Environmental and Economic Justice was created, partly out of frustration with mainstream environmental organizations "who choose not to accept leadership from people of color." It was also motivated by the need for "tools to be able to continue to advocate on our own behalf on environmental justice issues, whether they be in rural or urban settings, or on and off reservations and pueblos."[33] On their Facebook page today, they describe themselves as "Latinos/Chicanos, Asian Pacific Islanders, African Americans, Native/Indigenous and working class communities working for environmental and economic justice in the Southwest and Western United States and the Northern border states of Mexico." Also in 1990, Native people met at the Navajo reservation at Dilkon, Arizona, where the year before Navajo had succeeded in blocking a hazardous waste incinerator and landfill proposed for their community. They created a community group called CARE (Citizens against Ruining our Environment) and organized the first "Protecting Mother Earth: The Toxic Threat to Indian Land" conference, which was attended by over two hundred Native delegates from twenty-five tribes throughout North America.

Another significant event that year signaling the leadership of Indigenous people was the historic meeting in Quito of "nearly 400 Indian people, representing 120 nations, tribes and organizations of the Western Hemisphere."[34] Calling for a unified Indigenous response to the anticipated celebratory commemoration in 1992 of Columbus's arrival in the Americas, the Declaration of Quito reaffirmed that "We Indian Peoples consider it vital to defend and conserve our natural resources, which right now are being attacked by transnational corporations." Extending the critique to the "so-called powers," the nation-states that partnered with transnational corporations and "boasted of their development," the participants declared that the latter had only "deepened the level of inequality, ambition, crisis, ecological destruction" and "put the equilibrium of the planet in serious danger." They insisted on having a say in what happened to oil in their homelands, asserting that "with respect to strategic and non-renewable resources such as oil and uranium, the state may not negotiate in a isolated manner with a small local organization," but had to consult "the principal organization." If the state failed to do this, they declared, "these agreements will be considered invalid and non-existent." Participants also demanded "that national governments suspend indefinitely the authorization of permits to exploit renewable and non-renewable natural resources on our Indian territories."

Along with this critique of the dystopian, world-destroying dimensions of the collaboration between transnational corporations and nation-states in the service of resource extraction, the participants also declared their intention to "develop our own economic policies, based on the harmonic utilization of our natural resources, oriented primarily toward the betterment of our peoples and that will permit us to achieve shared ownership giving an alternative to the New International Economic Order." Participants also redefined "human rights," insisting on connections among human and nonhuman animals and the earth by arguing that in "consideration of human rights, we must take into account the continuous vital cycles of Mother Earth, air, water, the world, plants, those that fly, those that swim, the world of the four-legged, and all beings, because as human beings, we cannot live without all these natural beings."

World-making through activating Indigenous knowledge was also central to the 1991 First National People of Color Environmental Leadership Summit, where delegates learned about specific place-based environmental struggles such as the Havasupai Nation organizing against uranium mining in the Grand Canyon and organizing against Waste Management, Inc., the largest U.S. toxic waste disposal company, which disproportionately sited its hazardous waste facilities in close proximity to communities of color in places like Chicago's South Side and Port Arthur, Texas. The myopia about intersectionality in the mainstream environmental movement and its cooptation by some of the very forces it needed to fight were two of the biggest concerns of the summit. Delegates created an international policy group "in recognition of the global nature of the environmental crisis and the need for international cooperation to achieve solutions."[35] They also decided to present policy recommendations at the UN Conference on Environment and Development (UNCED) in June 1992 in Brazil. As well, the summit collectively produced a seventeen-point "Principles of Environmental Justice" document and a "Call to Action," which have become foundational for the environmental justice movement. In these documents, they called governmental acts of environmental injustice "a violation of international law, the Universal Declaration on Human Rights, and the United Nations Convention on Genocide." As well, they asserted that "environmental Justice must recognize a special legal and natural relationship of Native Peoples to the U.S. government through treaties, agreements, compacts, and covenants affirming sovereignty and self-determination."

That same year, in 1991, the second "Protecting Mother Earth" conference took place near Bear Butte, South Dakota, involving coalitions from Pine Ridge and Rosebud Reservation and co-sponsored by many other Indigenous organizations. It attracted over five hundred people from more than fifty tribes, who decided to form the Indigenous Environmental Network (IEN). IEN's activities include "building the capacity of Indigenous communities and tribal governments to develop mechanisms to protect our sacred sites, land, water, air, natural resources, health of both our people and all living things, and to build economically sustainable communities."[36] Echoing the language of the Declaration of Quito, one of the organization's "guiding principles" is that "as indigenous peoples, our consent and approval are necessary in all negotiations and activities that have direct and indirect impact on our lands, ecosystems, waters, other natural resources and our human bodies."

In June 1992, the Earth Summit, as the UN Conference on Environment and Development held in Rio de Janeiro came to be known, was especially important for Indigenous peoples and their rights related to the environment. It was a gigantic conference building on the first UN environmental conference twenty years earlier and included a final two-day meeting attended by delegates and heads of state and a parallel nongovernmental organization conference, Global Forum 92, attended by about eighteen thousand participants, including many Indigenous people. The conference produced the "Rio Declaration," a statement of broad principles to guide national conduct on environmental protection and development; treaties on climate change and biodiversity; a statement of forest principles; and "Agenda 21," a massive document presenting detailed plans for sustainable development.[37] Principle 22 of the Rio Declaration recognized that "Indigenous people and their communities and other local communities have a vital role in environmental management and development because of their knowledge and traditional practices."[38] As well, Agenda 21 included a section called "Strengthening the Role of Major Groups" that extended formal recognition of the rights and crucial roles of Indigenous peoples. Another important achievement was adoption of the Convention on Biological Diversity recognizing the significance of biological resources for Indigenous communities and the importance of traditional knowledge to conserve biological and species diversity.

In 1995, the Mole Lake Sokaogon Chippewa "became the first Wisconsin tribe to be granted independent authority by the Environmental Protection Agency (EPA) under the federal

Clean Water Act to regulate water quality on their reservation," including the Crandon mine that Exxon hoped to operate.[39] That same year, the Diné Community Action for Renewed Environment (CARE) organization became the first Indigenous group to make the Bureau of Indian Affairs produce an environmental impact statement (EIS), after many years of Peabody Coal and other corporations rubber-stamping EISs for projects on Native lands. And in 1998, IEN organized the Native Peoples/Native Homelands Climate Change Workshop in Albuquerque, which produced the "Albuquerque Declaration" sent to the UN Fourth Conference of the Parties of the UN Framework Convention on Climate Change. Since then, IEN has participated in climate change meetings on many different scales, including the local, regional, national, and international.

Today IEN has grown to an international coalition of over forty grassroots Indigenous environmental justice groups. Because "many of the struggles in Indian country are between Tribal governments and grassroots Indian groups, Tribal nations are not represented on the national council."[40] IEN holds an annual Protecting Mother Earth Conference to connect people by highlighting a particular local struggle. As a grassroots organization comprised of members of many different tribes and nations and "an activist group" that has a "strong spiritual component to its work and identity," IEN organizers specifically imagined and created "a network, a body that would share information among its members" (140) rather than a national organization. In Executive Director Tom Goldtooth's words, "IEN acts like the hub of a wheel, providing a common place through which individuals and communities can communicate," sharing information about similar struggles" (145–46). Something "that came out of our network," he further explains, "was deciphering all of this technology and these terms into language that our people understand" (146). IEN thereby models a powerful and illuminating engagement with science and technology that is a significant example of Indigenous futurisms reshaping climate change and striving to decolonize land and the imagination.

ANAMATA FUTURE NEWS OF CLIMATE CHANGE AND INDIGENOUS SURVIVANCE

While Native American and Indigenous science and futurisms have responded to struggles over the future of climate change at least since the 1970s, as Dillon shows in *Walking the Clouds,* such futurisms are also powerful in our recent past and present, animating social movement struggles and making deep connections across time and space. A particularly compelling example is *Anamata Future News,* a Māori web series delivering "breaking news, sport, technology, weather, and entertainment from the year 2018 til the year 2499," thereby showcasing "indigenous Māori language, arts, sports and culture in a futuristic world."[41] The Facebook page for the series calls it a "science fiction exploration into the future of Aotearoa"—the latter the Māori name for New Zealand. Each episode begins and ends with futuristic Māori news announcers greeting the audience in the Māori language. The series adeptly mixes humor and visually arresting sci-fi imagery as it explores "some of the big decisions New Zealand will make over the next five hundred years including energy use, climate, food, technology, crime, and the evolution of future generations."

The series imagines Māori people as part of the near and far future, beginning just a few years ahead, in 2018, and ending in 2499, as Māori encounter aliens and also begin to travel outside the galaxy. Throughout, Māori culture and politics thrive in the future, growing even more powerful over time. In 2135, the name of the country officially changes from New Zealand to Aotearoa as Māori successfully press for bringing back the "old names." But the series also imagines Māori as adopters of new technologies, users of social media, and producers of images, videos, music, and other forms of culture. One episode features a news story on how Apple's 2.8 OS can be inserted directly into the cornea, emphasizing that it offers GPS as well as MIRI 2.0, the Māori operating system. Mixing old and new, *Anamata Future News* brings older arts, such as carving and weaving, into the future, by combining them with new technologies in eye-catching sci-fi scenes. The past is co-present and the past, present, and future flow into each other in ways characteristic of Native slipstream.

Many episodes of *Anamata Future News* focus on climate change, resource extraction, and renewable energy. The weather map and forecast are featured prominently in each episode, with Co2 levels at new maximums and a warming Pacific Ocean producing a barrage of cyclones in 2025, and droughts continuing with freshwater sources dropping and scorching temperatures in 2040. Acid rain storms become increasingly frequent after a 2040 volcano-mining spill, which is not cleaned up until 2079, and by 2198 people are wearing 300 SPF sunscreen and no one can imagine eating anything that comes out of the toxic ocean. It's not all bad news, though, because farther ahead in the future Māori become world leaders in renewable technology because of an ancestor who imagined the possibility of solar forests. And by 2499, guerrilla gardens start to trend as some Māori build a movement around growing your own food here on Earth even as other Māori go beyond the known galaxy to explore the universe.

One of the key questions linking episodes is whether corporations should be able to mine the volcanoes in Aotearoa. In the first episode, set in 2018, Māori cast votes on phones, glasses, watches, and other devices in an election (#DigiVote) to decide if Aotearoa should transition to 100 percent renewable energy by 2040. In episode 2, set in 2025, power outages take place across much of Europe and Asia. Because other nations need help with global energy shortages, they have the majestic volcanoes of Aotearoa in their sights. We then learn the outcome of the election seven years ago: Aotearoa voted against renewable energy and now a second DigiVote will take place to decide whether to allow mining companies access to the country's rich volcanic resources. By 2040, the focus of episode 3, radioactive materials are leaking from a volcano mine and dolphins are endangered. The mining company Mega-Corp says the leak is contained but that people should stay out of the water until further notice. Finally, in 2075, Aotearoa's last untouched volcano becomes the site of a major environmental stand against resource extraction, as a land occupation springs up and protestors refuse to leave the site. Camping in front of the volcano, protestors wave signs bearing messages such as "I'm standing up for the land." Meanwhile, leaked documents reveal that the mining company has been buying digivotes and now has massive influence over government. In the face of such wastelanding by corporations and the state, the protestors insist they are here to stay and are not going anywhere.

The far future is brighter, despite ongoing climate change disaster. By 2198, there has been a catastrophic collapse of the volcanic-mining industry and Māori are emigrating to the Moon, where their kids enjoy antigravity playgrounds. Back on Earth, in the far future, there are still radiation storms but also cleaner water and because of an air filter forest made from a former rich people mega-tower, air quality becomes better for Māori on Earth. In the final episode, which takes place in 3499, the weatherperson announces that in some places there is no need for air masks today. In these ways, past, present, and future flow into each other in *Anamata Future News,* and speculative fiction and Indigenous futurisms become important methods for imagining a future of climate change in which Māori people not only survive but remix older forms of culture and knowledge with new technologies. Like the Standing Rock Sioux and other Indigenous people, the Māori thereby imagine and create other worlds in the face of imminent climate change disaster, disproportionately putting their bodies and lives on the line and also leading the way in the struggle for climate justice and people power.

TWO

Climate Refugees in the Greenhouse World

Archiving Global Warming with Octavia E. Butler

After the groundwater has been sucked out of the Ogallala Aquifer in *Almanac of the Dead,* Silko's character Sterling imagines that "great herds of buffalo" will return and only "those human beings who knew how to survive on the annual rainfall" will remain. By making Sterling foresee the depletion of the aquifer, one of the world's largest underground sources of freshwater, Silko responded to late 1980s news stories about humans draining and contaminating it. Silko thereby extrapolates from her present and moves backwards, forwards, and around in time to create a powerful Indigenous futurism in the face of ongoing battles over resource extraction and the wastelanding of Indigenous places in the U.S. Southwest and elsewhere in the Americas.

In 1990, one of these stories reached the ears of the late, great science fiction writer Octavia E. Butler, who carefully took notes on it after listening to an audiotape of a 1984 PBS episode of the TV show *Nova* entitled "Down on the Farm," focusing on the dangers of the aquifer's depletion.[1] A year later, as she was finishing her classic science fiction novel *Parable of the Sower*

(1993), one of the first to imagine possibilities in the wake of climate change disaster, she reflected on the episode's significance in a large spiral-bound notebook, one of dozens she kept from the time she was a teenager until her untimely death in 2006. In these notes, she regrets that the aquifer "is ½ gone, and the going has been done this century."[2]

A big part of the problem, Butler speculated, was that humans aimed to "maintain a familiar standard of living" even when it was destructive and would cause problems in the future. In California, she theorized, lawns exist because "non-Hispanic whites from the east—from wetter climates" recalled them and "wanted the living green fragrant mats as bits of the homes they'd left." Critical of human efforts to remake places they settled in destructive ways, Butler charged it was stupid, wasteful, and utterly without foresight—this last an especially significant insult coming from her—to hubristically transform the desert into lawns, golf courses, and "power-eating cities of light and night" such as Las Vegas, Laughlin, and Phoenix.

"They spend their tomorrows today" is a critique Butler leveled repeatedly at neoliberals who sacrifice the future for short-term gains and economic growth in the present, prioritizing immediate profits over water, the climate, and the earth. Butler's lament that "all we do is destructive" presupposes the value of mutualism over the kinds of parasitism encouraged by neoliberalism, which imagines a world made up of isolated individuals competing with each other to turn resources into property and extract profit in ways that supposedly are best for everyone. As Donna Haraway suggests, many biologists have also used "possessive individualism" as a template for understanding nature. Indeed, Butler's interest in "symbiogenetic imaginations and materialities" makes her a theorist of what Haraway calls a "New New Synthesis" in "transdisciplinary biology and arts" that moves away from modern science's rooting in "units and relations, especially competitive relations" to explore "symbiosis and collaborative entanglements," the "vast worldings of microbes, and exuberant critter bio-behavioral inter- and intra- reactions."[3]

Butler often used the language of symbiosis to think about human and nonhuman animals' future on the planet. In 1990, while writing *Parable of the Sower,* Butler acknowledged humans "are symbionts upon the earth" but that "not all symbionts are alike." Although humans were presently "parasites, destroying our environment," she hoped we could "become mutualists—symbionts who truly partner the earth, benefiting it as it benefits us," or at worst, doing no harm to it. Butler believed "parasitism upon our environment" was "not sane behavior" but rather "greed, shortsightedness, denial, self-indulgence, indifference, death."[4] She understood this deathly parasitism and shortsighted, greedy indifference to be a defining feature of the neoliberal political world around her. Connecting this struggle over the planet's future to contemporary political questions, she predicted that "those who know best how to conserve, restore, enrich, recycle, are likely to be defeated and overrun by those who know best how to take, exploit, and cast off. Easier to steal than to conserve—though oddly, the takers, the destroyers, are frequently known as 'conservatives.'" On the other hand, she observed, those who oppose "forms of exploitation that bring short term gain and long term desertification are called 'short-sighted' (!), 'against progress,'" and worse.[5]

In this chapter I tell the story of the emerging U.S. climate change conversation in neoliberal times by turning to Butler's *Parable* novels and what I call her memory work in assembling and organizing over the course of her lifetime a massive collection of clippings, reflections, and other writings, including many on climate change, the environment, and the destructive history of neoliberalism, especially during the Reagan era. Archivist Natalie Russell at the Huntington Library in San Marino, California, spent several years processing the collection, which ultimately filled more than 350 boxes, before opening it to researchers. Butler coined the word HistoFuturist to describe herself as a memory worker and "historian who extrapolates from the human past and present as well as the technological past and present."[6] In what follows, I argue that Butler's speculative archiving and imagining of worlds that were significant distortions of her present are connected and make her an important early climate change intellectual.

Butler's archive has recently inspired many people to produce projects in dialogue with her papers, including poetry, sound pieces, visual art, and theory. In 2016, the Los Angeles arts nonprofit Clockshop, founded and directed by Julia Meltzer, produced a "yearlong celebration of the life and work" of Butler entitled "Radio Imagination: Artists and Writers in the Archive of Octavia E. Butler," which included performances, film screenings, and literary events, including one I co-led with Ayana Jamieson, founder of the Octavia E. Butler Legacy Network, who is writing Butler's biography, and philosopher Amy Kind; as well as ten commissions by twelve contemporary artists and writers.[7] In this chapter, I join the conversation by emphasizing Butler's significance as a major climate change intellectual and by understanding Butler's collecting as a significant form of research or study.[8] As well, her archive reveals blind spots in discourses about the environment and climate change, partly through her angry annotations on newspaper and magazine articles and extensive notes on her research, appearing in notebooks and clippings in manila envelopes under headings such as "The Environment," "Science," and "Disaster." At a time when many despair that climate change science is too difficult for people without advanced science degrees to understand, Butler's critical archiving activity as well as her imaginings of forms of symbiosis beyond possessive individualism are especially illuminating.

Over the course of her lifetime, Butler carefully preserved and annotated hundreds of clippings about scientific research on changing weather, the greenhouse effect, and alternative forms of energy along with news of the destructive neoliberal political and economic policies that were precipitating ecological collapse. Butler started doing this research when she was just eighteen years old, only a few years after the publication of Carson's *Silent Spring.* Over the course of the next four decades, she speculated at length in journals, on notecards, and in her notebooks about what the future would be like as she struggled to imagine alternatives to the greenhouse world she saw emerging around her. She also linked climate change decisively to economic and political neoliberalism as she analyzed the coming catastrophe and the possibilities that might emerge in its wake.

Butler's climate change research was deeply entangled with her research on political and economic neoliberalism: the idea that individual liberty and freedom can best be protected and achieved by strong private property rights, free markets, and free trade. Butler was harshly critical of this worldview, raising sharp questions about how to shape climate change in the context of neoliberalism's cruel near future within which we currently live. Indeed, Butler's life as a writer and her archiving work span roughly the same years as most periodizations of neoliberalism: from the late 1960s or early 1970s to the present. In strong readings of Butler's 1990s novels, Mike Davis, Tom Moylan, and others have analyzed how her vision of a near future was built up out of elements of her present, notably the spaces around her, and anticipated many of the neoliberal transformations on the horizon. In the *Parable* series, a right-wing utopianism generates dystopian spaces of disaster, neglect, and everyday misery familiar from the last quarter century, from Los Angeles's 1992 uprisings to New Orleans, Detroit, and beyond. David Harvey tells us that neoliberalism is before all else "a theory of political economic practice"; its utopian dimension is to imagine that "human well-being can best be advanced by liberating human entrepreneurial freedoms and skills within an institutional framework characterized by strong private property rights, free markets, and free trade." Such right-wing utopian, "liberating" ideals justify the dystopian worlds in which more and more people now live.[9]

Butler's archive makes the concept of neoliberalism very concrete and material as early as the 1970s. That was the era when Butler performed "scut work," mostly done around Los Angeles's downtown Broadway Avenue as a temporary laborer.[10] She exulted at getting laid off because she hated her job and tried to make a living as a writer while making the downtown Los Angeles Public Library her office and second home. In the United States during Reagan's presidency, as in Britain under Thatcher, privatization, deregulation, tax cuts, budget cuts, and attacks on trade union and professional power carried the day in ways that still shape our present, and Butler responded to these changes both in her novels and through her memory work of archival preservation and organization.

RACE, COLONIALISM, AND THE ENVIRONMENT: IMAGINING OTHER WORLDS IN THE 1970S

Situating Butler's archive in relation to histories of not only neoliberalism and climate change but also science fiction from the late 1960s to the early 2000s also illuminates how the 1970s was a pivotal decade for people of color in science fiction. During that decade, the gatekeeping boundaries and uncomfortable position of being one of a few Black writers working in the field were often painful to Butler. This was true from the early days, even when she was having some of the most formative experiences in her career. Butler often said that an important event in her life as a writer was her attendance at the then-fledgling Clarion Science Fiction and Fantasy Writers Workshop, after being recommended for it by prominent science fiction writer Harlan Ellison, who had been impressed by her writing for the Open Door Workshop sponsored by the Screen Actor's Guild that he and others led for Black and Brown Los Angeles youth. When Butler traveled to Clarion, Pennsylvania, in the summer of 1970, she was leaving California for the first time, alone on a Greyhound bus. After she arrived, she was dismayed to discover there were very few other Black people in the town and none at all at the workshop itself, though Samuel Delany was an instructor one week. In a letter to her mother, she reported she was "surviving" though "it would be a little better if some of the other people in the workshop were Negro." To both her mother and her best friend back in LA, she implored, please "write me and prove there are still some Negroes somewhere in the world." She told her mother that there was only "one other nonwhite person" in the workshop, an "American Indian named Russell Bates."[11]

Bates was a young Kiowa man who was living in Andarko, Oklahoma when he was accepted to Clarion. He went on to coauthor an animated episode of *Star Trek* that won an Emmy, "How Sharper than a Serpent's Tooth." At Clarion, he and Butler became friends, exchanging letters after the workshop ended and sometimes getting together at science fiction conventions.[12]

Being the only Black student at Clarion at a time when the field was still mostly white and becoming friends with Bates at the workshop pushed Butler to reflect more on connections and differences between how Black people and Native Americans were positioned in the United States historically and at the time. In the journal she kept that summer, she wrote that she was "the token Negro," not in the way that "Russ was the token Indian because if he never told anyone he was Indian no one would ever know."[13] In the Clarion context, she felt her hypervisibility as a Black person was different from Bates's nonvisibility to those who had little previous contact with Native people and who might even misrecognize him as European. But Butler also recognized that this ignorant settler nonknowledge had pernicious power effects and contributed to myths of Native people as vanishing Americans. She wrote that Bates "had to take a lot more jokes than I did because no one has told our little white friends that Indians don't really like it any more than Negroes or Jews."[14] As late as 1978, Butler and Bates attended a science fiction convention together at the Los Angeles-area Leuzinger High School, where Bates was a guest of honor. The ephemera Butler kept are a small portal opening up onto a largely invisible history of Indigenous people's and people of color's participation in the field of science fiction in 1970s Los Angeles.

During that decade, Butler and Bates were living economically precarious lives, hoping to make it as science fiction writers when there were very few nonwhite people in the field. It was also in the 1970s that Butler's archiving of research materials on the environment, climate change, and science and technology intensified, including material referring to Native people's struggles over land and resources. In "The Environment" folder, Butler saved a 1971 *Los Angeles Times* editorial urging lands be returned to the Havasupai of Arizona as the National Park Service and the Sierra Club sought to incorporate Havasupai lands, thereby expanding the boundaries of the Grand Canyon Park. Butler's late 1970s "Subject Files" folder on Latin America includes several on Native people's struggles against colonizers and extractors, such as a 1977 *Los Angeles Times* article about Brazil, bearing the headline, "Amazon Land Boom Pushing Indians Out. Hunting Grounds, Traditions of 42 Tribes Threatened."[15] In her 1977–78 files on "Minorities," Butler also saved articles about Native people protesting the desecration of rock art sites, problems with federal policies towards American Indians, and efforts to save dying Indigenous languages.[16]

As well, Butler raised critical questions about white supremacy, Native people, and the genre of science fiction during this decade. In 1976, she wrote in her journal that often "whites write stories of prevailing over the natives of some new world they discover" and only rarely, "of the natives prevailing over them." She fantasized, on the other hand, about writing "a new beginning," an alternate history that would ask the question: "What would America have been like with European influence but without the continued contact—if the U.S. were somehow cut off as Greenland was at around 1650?" Butler speculated that "without a continued supply of white people," Natives would have been able "to keep more of their own land" and would have had "more time to replenish their numbers after smallpox." This would truly be "a new start," she mused.[17] Butler's 1970s journals often include such small-scale visionary fictions, which as Walidah Imarisha has defined the subgenre, are oriented "towards building newer, freer worlds" and are distinct "from the mainstream strain of science fiction, which most often reinforces dominant narratives of power."[18] In this case, Butler turns the dominant imperial-colonial genre of science fiction on its head and uses it to imagine a different history in which whites never colonized the Americas and Native people remained in control of most of their lands.

A little over a decade later, in the 1980s, when Butler began writing the first *Parable* novel, some early fragments focus on Earthseed refugees traveling to the stars and colonizing another world. At times, Butler even imagines Lauren Olamina as a Harriet Tubman struggling to save her people from a "stagnant, rotting Earth" by leading them to "a promised land among the stars." While outlining the novel's plot, Butler wrote in her notebook that "Black Americans are the only truly stateless people," since "no African country is our home," while in United States, "our special 'history' sets us apart." In the same notebook, Butler mused that "Only Olamina's way has any possibility of working for black people" and added, "This is the story of one woman's effort to stake her people free of white domination, their own defeatism, and the limits of Christianity."[19] Some of these early fragments imagine Black refugees in space, while others focus on Earthseed as a heterogeneous movement devoted to making inventions and discoveries intended for space flight and planetary colonization, such as the DiaPause technology that allows travelers to survive decades of starship travel.

In some versions, Olamina tells readers that the Earthseed project, the scattering of humanity among extrasolar worlds, was her response to the ecological crash she saw coming. Other fragments feature Olamina's children bitterly opposing her plans to go to the stars because they want to focus on helping Earth's people. In an earlier outline of the manuscript, entitled *God of Clay,* as well as in the published sequel, *Parable of the Talents,* Butler makes Olamina a much more problematic character in relation to her daughter, Larkin, who "ceases to believe in the Earthseed Destiny" and "focuses on the misery she sees among the free poor of her world," who are "victims of global warming and economic upheaval" and will not be helped by "extrasolar colonies."[20]

As early as 1986, Butler speculated about writing a "story taking place on a future Earth in Greenhouse Effect," where "ordinary people" struggle to "survive in new climatic conditions" in which "coasts" are "inundated" and worsening heat creates "deserts" in North America. What "extinctions and readjustments" would people face? What "societal changes" would follow? Butler wondered.[21] For several years, she debated the question of whether the first *Parable* novel should take place wholly on Earth or partly on other worlds and as late as October 1989, remained unsure how to begin.

As she struggled to see the shape of the first *Parable* novel, Butler wondered whether it might be possible to imagine the "World" as the main character and if so, whether she should focus on the "biological reaction of the world" to the people, depicting it as an "immune-transplant rejection."[22] Humans would have two basic ways of dealing with a truly living planet, she speculated: they could "learn to live with it in symbiosis" or they could "kill it" and "probably die with it."[23] Butler even drafted a section in which colonists debate the question of whether or not to create a bubble on a new world that will protect human life at the cost of "wiping out every living thing in the area." When they decide to go ahead and do so, "wherever the membrane covered, the native life died" and began to rot in its own particular way."[24]

That same year, Butler wrote in her notebook about her desire to "write long horribly or beautifully seductive novels about Humans of Earth becoming true mutualistic symbionts of other individual worlds."[25] In 2001, when Butler returned to this novel, she reflected that it should include "stories of mutualistic symbiosis as well as of parasitism and commensalism," where one organism benefits from the other without affecting it.[26] Butler was unable to imagine such a truly mutualistic symbiosis in a novel about the colonization of other worlds, however, and she never published a book in which her Earthseed refugees live among the stars.

Although the protagonist of *Parable of the Sower,* Lauren Olamina, longs for her Earthseed community to fulfill what she sees as its destiny to go into space, in the first novel they never do. In 1989, Butler wrote in her notebooks that it was "all right" if her characters did not "go to the stars in this book."[27] In November of that year, she reminded herself to instead emphasize "the ecological catastrophe waiting in the wings or already in progress" and how it might lead to "new restrictions, fewer cars (mainly corporate), few free people—and those suspect."[28] By 1991, Butler had definitely decided that the first *Parable* novel focused on "a young girl in a grim near-future, impoverished world undergoing global warming." Lauren Olamina travels north from her lower-class gated community of Robledo, which has been destroyed by arson, and walks up the abandoned California freeway "with multitudes of others," assembling "a moving community, hunting a place to settle."[29] In the end, they finally reach the Northern California coast, where Lauren's lover Bankole owns a small piece of land, only to discover that Bankole's sister and all the people who lived there have been killed. The novel concludes with a funeral. At the end of the 2000 sequel, *Parable of the Talents,* on the other hand, the community finally does take off for the stars, ominously enough in a ship called the *Christopher Columbus,* which suggestively warns of the dangers of seeking "empire" and transposing the tragic history of European colonialism in the Americas into space.[30]

I understand the unpublished fragments, blueprints, and drafts of these prequels and sequels as a kind of dreamwork or unconscious relative to Butler's published novels. They illuminate paths not taken, other possibilities, and knotty problems that could not be resolved to Butler's satisfaction. In the years that followed, Butler intermittently tried to make progress on a third novel, *Parable of the Trickster,* writing pages and pages of notes, fragments, and short incomplete drafts until she finally gave up and turned her attention to her last published vampire novel, *Fledgling.* In March 2000, she called *Trickster* her protagonist's "Struggle to defeat Quick and Dirty ecological methods, corruption, and tyranny in her colony's partnership with one another and with their new world."[31] But just a few weeks later she worried in her journal that *Trickster* might not be salvageable and in the months before she died, on the back of a piece of stationery for the cardiovascular department of a Seattle medical clinic, she wondered how to make the novel "whole and sexy and strong."[32] At the end of her life, however, she hoped to return to the series after finishing a sequel to *Fledgling* and even imagined she might be able to write four more Earthseed novels *(Trickster, Teacher, Chaos, Clay)* about "four generations on a new world," culminating in the story of the "non-humans who have developed from the human colonists."[33]

As Butler tries to imagine humans leaving Earth behind, she keeps stalling on the problem of whether Native life exists on other planets and what it means for humans to impose their own way of life on other worlds. The latest extracts and drafts she saved for the archive dwell on the dangers and catastrophes that arise when humans begin to awaken each other after being placed in suspended animation for over one hundred years on a starship journey to settle new worlds. In some versions, her protagonist, Imara, is the community's archivist, librarian, and historian, while in others she is a cop, therapist, physician, scientist, or college professor who is a "female Stephen Gould"—"bright, tough, kind."[34] Butler situates humans in space but imagines them "fighting the misery of living for the rest of their lives on a world not their own," where "nothing at all is right." For a time, she played with the title *Xenograft,* rewriting the story as one of a doctor trying to help an extrasolar colony survive its infancy in the face of a disease called Graft-v-Host (GVH) that brings death to a third of those who contract it. Here, Butler extrapolates from James Lovelock's and Lynn Margulis's Gaia Hypothesis in a story of space colonists who "have torn ourselves away from the living tissue—the biota of earth—and (xeno)transplanted ourselves to a new world."[35]

While "striving toward utopias" and "trying as hard as they can to make societies that work," Butler's space colonists experience hallucinations and accidents, retreat into all-consuming religious practices, commit murder and suicide, walk away from the colony, go blind, and experience telepathy plagues.[36] Butler speculates that one faction might try to conquer the planet and seed the world with terrestrial plants while a second group "accepts the world and tries to learn more about it, partner it."[37] In *Trickster* fragments and plot brainstorming, Butler connected the story directly to the Reagan administration's dumping of "environmentalism, aid to education, health care," and "racial equality" in favor of "grabbing money, squeezing the environment for the last penny, screw trees and animals, water and air, people and progress," reflecting that "on another world this is terrifyingly deadly in such an unforgiving environment."[38] Wondering briefly if her space colonists' experience might be like the settling of the United States if there had been no Native people, in the next sentence Butler acknowledged that settlers "would have died without Indians to help them learn to grow new crops" and thus the comparison did not work.[39] In some extracts, the settlers only imagine there is native life while in others "real natives show up," who look "not in any way ghostly or immaterial."[40] In all of these ways, intellectual, practical, and ethical questions about colonization and the difficulties of creating a truly mutualistic symbiosis between people from Earth and other worlds seemingly prevented Butler from finishing a story in which Earthseed refugees successfully spread throughout the galaxies. In her published visionary Earthseed *Parables,* except for the last novel's final chapter, Lauren and her people remain on a ruined Earth, looking for possibilities in the wake of the climate change disaster that Butler's research told her was already happening.

ARCHIVING GLOBAL WARMING AND CLIMATE CHANGE IN NEOLIBERAL TIMES

While most of the essential research proving that climate change was spurred by human activity had been completed by the 1960s, only when the mean global temperature began to rise in the late 1980s did climate change emerge as the subject of widespread concern in the scientific community and for social movements. Butler saved materials as early as 1965, but there is a significant uptick in her clippings in 1981 and 1982, when Reagan was elected and began to make dramatic changes in U.S. environmental policies, and there are clusters of material about climate and the greenhouse effect in the late 1980s and 1990s, the years she was writing the *Parable* novels. Although earlier the state had responded to the emerging environmental movement by establishing new regulations, laws, and institutions, by the end of the 1980s, the Reagan administration rolled back and otherwise undermined many of them. Working on the neoliberal premises that governments should get out of the way of businesses and corporations and that citizens needed to be liberated from the nanny state to realize their potential as individual entrepreneurs, Reagan and his team were hostile to an environmental movement that warned of the costs to the planet and its people of unfettered capitalism and sought not only to regulate and control but also to imagine a collective good and create the solidarities and alliances necessary to realize it.

Butler's *Parable* novels introduce a core concern that deepens in her work over the course of time: ecological collapse and climate change, which she predicted would be major world problems. Butler understood both precisely as global forces whose damage would not be confined to a single part of the world though she also imagined that the worst of their impact would be unequally distributed and would hit the poor hardest. Butler's "Disaster" files include material from 1965 to 2005, including stories about tornadoes, hurricanes, floods, earthquakes, droughts, and fires, along with manifestly human-made disasters such as the 1995 Oklahoma City bombing and the 1992 Rodney King verdict.[41] "The Environment" folders, on the other hand, include dozens of newspaper and magazine articles published between 1965 and 2004 that overwhelmingly stress human agency in bringing about the catastrophic climate changes on the horizon. Butler saved a 1988 article about ozone depletion that emphasized that its protection must be "international responsibility" as well as early 1990s stories about the coming extinction of millions of animals and the necessity of taking "prompt action to curb global warming." Late 1990s articles track melting Antarctic glaciers that "could flood coastal areas, scientists say," or "link" to "global warming" storms such as those so common in her "Disaster" files. Explaining the purpose of this research in an interview in 1993, Butler called herself a "news junkie who can't help wondering what the environmental and economic stupidities of the '80s and '90s might lead to."[42]

During the 1970s, Butler saved articles on "disastrous changes in the weather around the world" and on a "somber report on climatic change and global food production" issued by the National Academy of Science warning there was "general agreement that more erratic weather will make it difficult to sustain the consistently high yields of food" of the 1960s and 1970s.[43] She also kept a clipping on "the Aerosol Threat," that emphasized the dangers chlorofluorocarbons (CFCs) posed to the ozone layer. Many of these clippings focus on solar and other forms of renewable energy, including a 1977 *Los Angeles Times* piece on an innovative solar housing project in Davis, California.[44] That same year, Butler carefully preserved stories about how climate change was affecting California, including one criticizing the water-wasting southern Orange County Mission Viejo project in a context of extreme "drought emergency" in "what could be the driest year in the state's history."[45] Another story looked to solar as a form of renewable energy that could solve Southern California's energy problems in light of "the failure of the 'invisible hand' of the free market to govern successfully the national production and use of energy," which the author considered a "fault imposed on society."[46]

One context for these 1970s stories was what one writer called "the quest for alternatives to expensive, embargo-prone Arab oil."[47] Another 1970s story focuses on how "ecology" has become a new "industry" in the wake of new state and federal laws mandating environmental impact reports so that decision-makers must think twice before "plowing [the environment] under for progress." Many of the articles, however, even as early as the 1970s, warn of the "drastic warming" of the "climate," such as a story about the National Academy of Science report, which "reinforces some of the more dire warnings of those who worry about the greenhouse effect," stating that "the possible climatic consequences of relying on fossil fuels may be so severe as to leave no other choice" but to press "for the development of alternative energy sources."[48]

During the 1980s, many of Butler's climate change-related clippings continued to focus on the greenhouse effect, but there are also several on the political dimensions of the climate change crisis, especially on the damaging policies of the Reagan administration. Butler was moved to write annotations in red ink on a 1981 editorial published in the *Los Angeles Times* introducing the Reagan Administration's "Environmental Swat Team," led by Secretary of the Interior James Watt, whose main claim to fame before being appointed was representing "power companies, oil and gas interests, ranchers, and miners in the Sagebrush Rebellion against regulation of the development of Western resources." The piece refers to an important historical struggle that took place after Nevada passed a 1979 bill taking control of certain lands within its boundaries under the administration of the Bureau of Land Management. Several other Western states did the same, inaugurating conflict among states, landowners, and the federal government over jurisdiction and lighting a fire under environmentalists, who worried public lands would be sold off and despoiled after their resources were extracted. Butler underlined in red a sentence stating that Watt's "grand design is to open up more 'locked up' government land to logging, mining, and petroleum development." She also scribbled "Profit in haste; Repent at Leisure" in the margins. As well, she underlined several sentences that emphasized the problems these policies would cause for people in the future, and wrote in large letters at the bottom of the page: "These are truly people who see the trouble they cause their children and grandchildren to be irrelevant."

The connections Butler made between the Reagan administration's neoliberal policies and imminent danger to the environment are also preserved in an *LA Times* editorial entitled "Nation's Leaders Must Remember Times to Come." Butler wrote "Environment" on the top along with the comment: "But instead they will sell our birthright for a quick profit." The story worried that Watt would make more Western lands available for oil drilling and strip mining, and asked "How long can we, in our arrogance, force the environment to meet our desires instead of fitting our desires to the environment"? In the margins, Butler wrote "Good Common Sense," after this question and added several comments in red ink a few months later, including her fear that "Watt's 'Fuck Tomorrow' attitude" will "destroy us." She also wrote angry red annotations on other articles about Reagan undermining the Air Quality Control Act and seeking to reduce "requirements on industry." Another article focuses on Watt's reopening of four California offshore basins for oil exploration leases and his opponents' criticisms of the Reagan administration's "almost religious zeal" in "promoting oil and gas development." By saving these 1980s clippings, Butler linked Reagan's effort to roll back new, post-1970 environmental regulations while opening up lands to oil, coal, and gas extraction with concerns over global warming and ozone depletion. Butler observed that almost every person Reagan appointed to watch over a department wanted to destroy it, including Ann Gorsuch, the first female administrator of the EPA and mother of Trump' Supreme Court appointee Adam. In a 1982 notebook entry Butler showed her disdain for female neoliberals, writing: "Margaret Thatcher/ Ann Gorsuch females are deadly to the species."[49]

While attacks on environmental regulations intensified during the 1980s, new scientific research on ice samples and deep ocean histories impacted conversations about greenhouse gases and climate change. Butler carefully documented this science news while questioning the Reagan administration's ecological obtuseness and idealization of short-term profit. At the end of the 1980s, Butler saved articles about a French and Soviet study that used 160,000-year-old ice samples to provide "the strongest evidence yet to link an increase in atmospheric carbon dioxide to warming of the earth—the potentially catastrophic 'greenhouse effect.'" In 1979–80, the article stated, "studies of Antarctic ice indicated that carbon dioxide levels had increased about 40 to 50% as the last glaciers retreated about 10,000 years ago."[50] Butler also preserved a 1987 article titled "Antarctica's Ozone Shield Shrinks to Thinnest Ever," about government scientists' announcement that the previous month the ozone had grown thinner "than at any time previously recorded." In 1988, she added an editorial about "the startling discovery of a significant worldwide depletion of the Earth's protective ozone layer," which insisted addressing this problem must be an international responsibility and called for a phase-out of harmful chlorofluorocarbons.[51] Finally, in 1989, Butler annotated another article about how global warming would create super storms like Hurricane Hugo, which that year caused fifty deaths, left one hundred thousand people homeless, and was the most expensive storm up to that point to hit the United States. Butler carefully underlined in green sentences that explained how a warmer ocean causes more evaporation and that warmer air can hold more water vapor, both of which increase the power of hurricanes. She also underlined the article's warning that warming ocean and air temperatures will increase wind speeds 20 to 25 percent and their maximum intensity by as much as 60 percent.[52]

Butler's creative collecting reveals how "climate" was increasingly imagined as a world problem during these years. Many of her 1990s clippings focus on major international meetings and agreements of the era regarding climate, including one about the 1990 Sundsvall, Sweden conference where the newly created International Governmental Panel on Climate Change (IGCC) released its First Assessment Report on global climate change. "Panel Warns of Disasters from Global Warming," the headline ominously declared, while the story reported on the "75-nation conference of scientists and government officials" that hoped to "set the stage for an international effort to combat pollutants that accelerate nature's 'greenhouse effect.'" Conference participants warned of "the potentially disastrous human and economic impact of global warming in generations ahead" and predicted that if no action was taken to curb emissions, "expanding seas and melting polar ice might produce sea level increase of as much as three feet by the year 2100," dramatically affecting 224,000 miles of coastline. That would render some island countries uninhabitable, displace tens of millions of people, threaten low-lying urban areas, flood productive land, and contaminate water supplies. The article also states that the United States "has resisted the idea of mandatory limits on carbon dioxide emissions," putting it "at odds with European nations, several of which have already set targets and deadlines for restricting greenhouse gases."[53] The memory work that Butler did thus emphasized both the efforts to forge international agreements and how the United States often stood in the way because of its prioritization of economic growth, corporate profits, privatization, and other neoliberal values.

Other articles in Butler's archive include stories about how in 1991 the National Academy of Sciences declared that the United States could cut greenhouse gas production by 40 percent with little economic cost, reversing a "hands-off recommendation" made eight years ago and calling for "a far more aggressive posture" in confronting this problem than the George H.W. Bush administration eventually pursued. Warning that Earth's average temperature could rise 9 degrees Fahrenheit, "with potentially catastrophic results in the next century," the article faulted Bush for refusing to adopt targets for limiting carbon dioxide as much of the rest of the world had.[54] Butler also kept stories about how global warming was affecting the ocean, causing whole populations of sea creatures to migrate northward, along with a 1992 article predicting that as many as half of the planet's species could vanish or dwindle to nothing in the near future, with large animals still existing only in scattered preserves or zoos, their survival dependent on frozen embryos, sperm, and eggs.

As well, Butler preserved a 1992 special section of the *LA Times* entitled "A Day in the Life of Mother Earth: A Special Earth Summit Issue of World Report," so carefully that it still looks pristine. This historic meeting was the largest environmental conference ever held up to that time. The cover warned that on a typical day the earth experiences 250,000 people being added to its population; up to 140 species become doomed to extinction; nearly 140,00 new vehicles join the road; deforestation of areas one-third the size of LA occurs; and "more than 12,000 barrels of crude oil will be spilled into the world's oceans."[55] Butler also saved a 1998 article entitled "Melting Antarctica Glacier Could Flood Coastal Areas, Scientists Say," which reported "Satellite images taken from 1992 to 1996 show the glacier is shrinking" and warned, "if it melted it could lead to the collapse of the West Antarctica Ice Sheet, causing global sea levels to rise as much as 20 feet."[56]

Butler documented most of the major flashpoints in the emergence of climate change as a world problem, driven by scientific research, social movements, and top-down neoliberal political and economic changes from the late 1960s through the early 2000s. Thus it is not surprising that in 1999 she also preserved an editorial called "Indifferent to a Planet in Pain," which Bill McKibben authored just as a new edition of his classic book *The End of Nature* was about to be published.[57] Reflecting on the decade since the book's publication, McKibben remembered how ten years earlier global warming had only been a "strong hypothesis." Now, however, "after a decade of intensive research," McKibben stated confidently, "scientists around the world" had formed an "iron-clad consensus that we are heating the planet." McKibben listed an array of disturbing facts in support of this consensus, including that spring now came earlier in the Northern hemisphere than thirty years before; that severe rainstorms had increased by over 20 percent, the result of warming air carrying more water vapor; that the Arctic ice sheet was in many places forty inches thinner; and that "warmer waters have bleached coral reefs around the globe," "glaciers are melting," and "sea levels are rising." Concluding "it's far too late to stop global warming," McKibben advised the best humans could do now was to "slow it down" through "stiff increases" to the price of fossil fuels, by supporting and incentivizing more research on renewable energy technologies, and raising fuel efficiency standards. Worrying "we don't yet feel viscerally the wrongness of what we're doing," McKibben argued people needed a "gut understanding" of "our environmental situation" if they were "going to take the giant steps" that they "must take soon." In order to envision what life would be like a hundred years from now and make the necessary changes, McKibben concluded, people needed to feel and understand on a bodily level that "we live on a new, poorer, simpler planet and we continue to impoverish it with every ounce of oil and pound of coal that we burn."

CRITICAL DYSTOPIA IN THE FACE OF DISASTER

The dry, harsh, austere world too poor for lights, cell phones, public schools, and other things still essential to our world, and where water is a luxury to which the poor have only intermittent access looms ominously at the beginning of *Parable of the Sower*, which provides a visceral imagining of life on a poorer, simpler planet, the greenhouse world created by human-produced global warming. Perhaps giving her readers a gut understanding of life in such a near future was what she had in mind when Butler created large, color-coded notes exhorting herself to add "More Heat & dust & thirst & stench & misery & Fire" and "Show the Greenhouse World HOT POOR DRY—or Drowned or Blown Away or BURNED." Butler wrote that she was partly reflecting on conversations with close friends about imminent ecological disaster: "Consider: Olamina must live in the hellish world that Donna, Victoria, Maggie, and I have discussed." Maggie "sees economic devastation for all of America and lack of resources as a result of the greenhouse effect, acid rain, and other ecological problems."[58]

This is the world Butler's protagonist Lauren inhabits at the outset of her novel, "an un-privileged enclave" as Butler called it in a 1988 notebook entry.[59] In the opening, Lauren dwells inside the ill-fated gated community with her parents and brothers, before fire is the phoenix that burns out the old way of life and forces Lauren to search for possibilities in the wake of disaster. As she moves north up abandoned highways, dressed as a man, she collects people to form a community of disposable, beaten down, vulnerable folks of many different races and national origins. Avoiding the transnational corporations ready to exploit especially white workers willing to endure new forms of slavery, Butler's roving, multiracial enclave of resistance knows better than to put its faith in nation-states or monsters of corporate capitalism. In 1990, Butler described the novel as "the coming-of-age story of a woman's struggle to help the free poor of her time to unite, help one another and partner the Earth—restoring it, not subduing it."[60]

Elsewhere in the archive Butler clarifies the stakes of writing these speculative fictions of climate change.[61] On a notecard for a speech, she wrote: "My novel shows us a society that did not prepare itself to deal with Global warming because the warming isn't just an incident like a fire, a flood, or an earthquake." Instead, she suggested, "it is an ongoing trend—boring, lasting, deadly—that feeds on itself." Even though it would not happen all of a sudden, however, Butler predicted that global warming would have dire effects: food and water prices would go up, sea level rise and greater coastal erosion would threaten places such as the Bahamas and Florida, and there would be severe drought, fire, and storms. "We can't avoid it and we aren't preparing for it," she worried, fearing the addition of climate change to all the "usual stuff," including "racism" (which she crossed out), "earthquakes, social turmoil, etc."[62]

In earlier drafts, Butler was much more explicit about global warming, naming it as the precipitating cause of this harsh, austere, near-future world. "The country, the world, was in rapid, indisputable transition," she wrote in an early fragment, because "climate change was global, indisputable, and ongoing." Southern California is hit especially hard, its problems with water growing dramatically worse. Politicians finally begin to make laws restricting the burning of fossil fuels, but it is too late: "Climate change was well advanced and would not begin to retreat during the lifetime of anyone now suffering," even though people have been "frightened into" curtailing driving and other uses of "fossil-fuel generated power" (66). Although "obvious water and power hogs" such as Phoenix and Las Vegas suffer first, residents of Los Angeles, Orange County, and San Diego no longer have air conditioning or much water. Global warming explains why in the draft Butler's characters are walking from LA's Baldwin Hills down to Wilshire Boulevard, which is over four miles away. It also explains, as Butler put it, "why they are having so much difficulty with water." The heavily guarded water stations and itinerant water merchants who sell this increasingly precious and rare commodity to the roving poor were inspired by Butler's rigorous extrapolation from ominous trends in her present.[63]

In a 1989 fragment, she expanded on the history of the "drought that never ended," as Lauren's parents' generation thought of it. Even before she was born, Lauren recalls in a part Butler ended up cutting, "government, business, and agriculture" were forced to confront climate change disaster and so, in "grudging, uneven, uncertain, inadequate fashion, they dealt with shortages, crop failures, transportation problems as river and lake levels dropped in some parts of the country and sea level rose in others." Inflation intensified, driven by food and fuel prices as people began to be charged "according to how much the fuel they bought would contribute to the ongoing global warming." Still "things changed slowly" and the situation had "become almost normal." People had known what was happening for a long time, Lauren remembers: she "read books about it from the 1980s and 1990s and they refer to even earlier books." Accurately predicting what science fiction writer Kim Stanley Robinson has named "The Dithering" over climate change that would ensue in the years after Butler's death, Lauren rhetorically asks "What did they do in response?" before supplying the answer: "Not enough."

In other unpublished fragments, Butler brainstorms about how climate change might shape class divisions in the future, speculating about the privatization of public education and the need for poor communities to produce their own foods by creating urban gardens such as those that Detroit activists Grace Lee Boggs and adrienne maree brown helped to create. While some people with money and connections in Butler's *Parable* are able to live in luxury as predicted by the futurisms of "late twentieth-century books," Lauren tells us, people like her father Lawrence, "a teacher at Robledo City College, still working in the public sector, were not paid enough to live." Because he couldn't support his family as a college professor, in one early variant Lawrence, who is Black and Mexican in several drafts, quits his job to work as a greenhouse gardener in nearby walled and gated communities. These gardens "were not in greenhouses or shelters of any kind," Lauren clarifies, but rather were "gardens grown to supply the table when markets offered little or charged outrageous prices for simple produce or canned goods." The "sparse living" he earns as a gardener, along with the school-keeping of his wife, Cory, situates the Olamina family "among the more affluent of the working poor," while only "Lauren and her stepmother Cory could remember" when Lawrence "had taught history and political science at the local community college." In this variant, a corporation takes over the college, cuts wages, and fires people in what Lauren presents as a familiar pattern of neoliberal privatization: "Every time a government-run institution went private, wages dropped, and some people lost their jobs." The company asks Lawrence to stay but he walks away because he would be required to teach "bad history," including that "Asian, African, and old Soviet disregard for the environment"—but not that of the United States—"caused the worldwide warming trend that had brought so much scarcity." The corporation also wants professors to teach the lesson that "large numbers of Hispanic, Asian, and African Americans" have "little claim on the United States" and should therefore "be examined and expelled from the country if they could not prove their claims of citizenship." Lawrence does not want his children to grow up believing those lies and to become part of "a block of loyal, miseducated people," brainwashed by the company's network of schools, businesses, and housing, "who would be useful" in "any efforts it intended to control local, state, or even national government."[64] In other early fragments, Butler's Earthseed community operates desalination plants on the principle that "water's life, what's more precious than that?" (54) and parades of poor people march to protest water quality failures.[65]

Butler drew on her extensive research files on refugees and displaced people, notably including several articles about the resettlement of Vietnamese refugees in Camp Pendleton, California, in imagining Lauren as a kind of refugee. These materials prompted her to think about how migration and demographic changes perpetuated new forms of slavery and disposable labor around the world. Her *Parable* novels are full of characters who have been forced to endure various kinds of slavery, from indentured labor to sex slavery to border-works where companies take advantage of the deregulated zone to push workers ever harder and authorize the worst abuses, in a significant distortion of labor conditions in the U.S.-Mexico borderlands in the wake of the 1993 North American Free Trade Agreement (NAFTA). Since Butler dreamed up Lauren Olamina over a quarter century ago, her concept of "climate refugees" has made its way into climate change discourse, partly because "predictions for climate change-induced displacement" range "from 150 to 300 million people by 2050, with low-income countries having the far largest burden of disaster-induced migration, according to the Internal Displacement Monitoring Center."[66]

When Butler called her protagonist Lauren Olamina a refugee back in the early 1990s, she had no illusions about whether humans had caused climate change: she was sure they had and were. She includes many anguished passages in her journals and notebooks on the lack of foresight of U.S. political leaders after Reagan in dealing with the imminent disaster of climate change. Reflecting in 1991 on George H. W. Bush's response, for instance, Butler wrote in obvious frustration that "his reply to people worried about global warming is, "Don't worry, it isn't going to happen." Because he was clueless about climatology and because it "threatens something he does know," she rightly feared, he would not take action to prevent a climate change dystopia. If "he lives to see any of it—flooding, drought, economic mayhem, etc.," she hypothesized, "he will not perceive his own responsibility for it."[67] On May 29, 1992, in the wake of the LA uprisings of the previous month, as she was finishing *Parable of the Sower,* she speculated that "maybe the riots, the anti-green attitudes of the administration, and the general decline (economy, ecology, ethnic relations, etc) deepen my gloom, but that's all except the riot been around for a while."[68]

In keeping with the neoliberal accord between Democrats and Republicans despite Democrats' greater stated concern for the environment, the Clinton years (1993–2001) produced more attacks on welfare and the homeless as well as the establishment of NAFTA, which allowed corporations more freedom and flexibility than persons in crossing geopolitical borders. NAFTA also helped create infamously exploitative *maquiladoras* operating in environmentally reckless ways in borderland zones. Meanwhile, privatization continued apace and nation-states and fossil fuel interests failed to imagine alternatives to climate crisis and make necessary changes. This led Butler to predict that the 2020s would be the decade of collapse in which humans would witness sea level rise, dryness, heat, crop failures, institutions no longer working or existing only to collect taxes and fees and to arrest people to exploit their labor: "This is the story of 'The Burn,'" she wrote, "a period in history when old ways of life were dying as the climate changed, food and water became scarce, expensive, unsafe, and the focus of much criminal activity and new ways were being born." Calling *Parable* the "story of one woman who builds her 'new way' upon the ashes of the old," she imagined the near-future 2020s as "the Burn" and the Earthseed community as "the head of the Phoenix, rising."[69] In other words, Butler imagined neoliberal globalization from above as a kind of scorched earth disaster, one to which her imaginings of different worlds and communities and other, more sustainable ways of living responded.

Peering ahead into the near future, Butler saw the decade beginning in 2000 as a time of confused recovery from the 1990s, except that the "ecology does not truly improve. There is too much heat, too little water in one place and too much flooding somewhere else, crops do poorly" and "eco-hardship becomes a way of life while there is still money in denying" the catastrophic changes that are taking place. In the 2010s humans are still ignoring the "ecological holocaust," which means "Hell is at hand." By the 2020s, everything has to change and the old ways have to die, even as people "wall themselves in and suffer privatization." Even worse, "the seas are rising" and the "air is hot and dusty and brown." In the 2030s ecological changes cause truly hard times, with only a few rich people doing well: "Walled in, flying, boating, and tanking in supplies," they take on poor people as indentured servants in return for room and board. By 2040, companies have been given broader rights to continue the "slow environmental degradation" of "rising seas, searing deserts," cancer, unbearable heat, and more, while religious authorities promise "if we right ourselves" we can return to the old days "when God liked us best." In imagining the future, Butler kept returning to climate change disaster in decade after decade.

In speculative notes written at the end of the century in her journal, in response to her own question "What Would I Like the World to Be Like?" she wrote: "More solar, wind, water, and other renewable-energy sources put to use. Less fossil fuel use."[70] Emphasizing that "global warming is real" and it's going to cause "much misery, much suffering," she prophesied that the world would face slow violence, otherwise described as "a series of chronic problems, not a big, acute crisis," and that "we'll be dealing with it for a long long time—far longer than anyone alive now will live." Worrying over the "health aspects of global warming" as "tropical diseases move north" and "sewers in coastal areas" are invaded by sea water," Butler also predicted problems with "supplying populations with potable water," and the necessity of working harder "to prevent insect-vectored disease." Thus she wrote she was glad to see the August 1999 issue of *Science News* explore "some aspects of learning to live with global warming, since it is too late to prevent it" and "it will affect every human being on Earth in more ways than we yet realize."

Instead of "storing up future disaster for immediate wealth" as neoliberals around the world did, Butler advised, we might instead start preparing people for the climate changes to come, partly by changing the ways we educate. "A school that teaches questioning and problem-solving must prepare people to deal with global warming," she mused in early nineties speculative notes. By significantly distorting her present to imagine a different future, Butler rigorously kept track of how even as the "old ways of life were dying" due to climate change, "new ways were being born." By imagining a community that makes "new ways" in "the ashes of the old" and by critically documenting the intersecting histories of climate change science and neoliberal politics from the 1960s through the early 2000s, Butler's visionary fiction and HistoFuturist archiving continue to illuminate the new inequalities, divisions, and solidarities created by climate change.

THREE

Climate Change as a World Problem

Shaping Change in the Wake of Disaster

In *Emergent Strategy: Shaping Change, Changing Worlds,* activist and speculative fiction writer adrienne maree brown advises that in confronting climate change today, "we need to have a level of dystopian consideration": "Certain climate realities are no longer wild imaginings, they are happening, and they are coming. (*Game of Thrones* watchers? Winter is here, and it's balmy.)." Referencing the HBO show based on George R. R. Martin's fantasy novels to make the point that climate change is already happening, brown explains that in this context, "Octavia Butler appeals to me because she wanted to prepare us" for the changes that are now inevitable: "Change is coming—what do we need to imagine as we prepare for it?"[1]

In responding to imminent climate change dystopia it is useful to have a theory of shaping change, which brown has discovered in Butler's work. Brown's ideas about shaping change in social movements were especially inspired by *Parable of the Sower.* Indeed, brown claims that Butler is "one of the cornerstones of my awareness of emergent strategy" (6), which is partly based on the leadership model found in Butler's fiction. Brown suggests that Butler "created case studies that teach how to lead inside of change, shaping change" in ways that are "relational, adaptive, fractal, interdependent, decentralized, transformative" (56). Brown applies these ideas in facilitation and organizational development work, much of it involving climate change, which she has been doing since the early 2000s.

I first became aware of brown in 2010, when Amy Goodman interviewed her on *Democracy Now* about the U.S. Social Forum in Detroit, attended by over ten thousand people from more than one hundred grassroots organizations. Brown was national co-coordinator of the program, organizing coalitions by using digital and media technologies as well as speculative fiction and Butler's novels. She was also director of Allied Media Projects in Detroit. The U.S. Social Forums had begun in 2007 in Atlanta, inspired by the World Social Forums that had emerged as a kind of grassroots globalization movement a little earlier. In the interview, brown explained it was "very groundbreaking" to "say we're part of an international community trying to actually hold accountable our corporations and our governments," which have international as well as local impacts. She pointed out that the World Forums outside the United States responded to the elitist, top-down nature of the World Economic Forums at Davos, Switzerland and other summits "where basically all of the big money folks would get together around the world and say this is what we think the solutions are." She found inspiration in how World Forums outside the United States used culture broadly conceived, including film festivals, performances, and people's assemblies, to imagine different futures of climate change and social justice. Calling ecological and climate justice central to the forum's work, brown told Goodman she planned to create a People's Movement Assembly for organizers who have "been working on the relationship between people and planet, to come together and say what are the major priority issues." She led a protest on the forum's closing day in which thousands of activists marched and rallied at the Detroit incinerator, the largest in the United States, demanding its closure. Brown was especially excited by how ideas about creating change moved through the workshops and into the People's Movement Assemblies, then into action and into the streets.

At the same time, brown was also co-creating materialist Black feminist futurisms through media justice work and by using Butler's speculative fiction as a springboard for organizing communities. In the interview, she talks about her involvement with Allied Media Conference, a gathering of "visionary media makers," operating on the premise that "everyone can be a media maker," teaching "folks to make their own radios, make their own computers, and really be the people who tell the story and create the history of their own communities." At the conference, brown explained, she facilitated an Octavia Butler symposium where participants talked about postapocalyptic survival and how Butler's ideas were "relevant to us now as organizers who are in the throes of change that we don't have any capacity to comprehend." It's "easy to be devastated," she observes, but "I think Octavia Butler's work calls us to be inspired instead." Inspired by Butler, she has worked with many movements, doing environmental, food, reproductive, gender, economic, and other justice work, and collaborating with organizations that "do harm reduction work with active drug users and sex workers, voter organizing at the national level, food justice work in Detroit, and nonviolent direct-action training, primarily supporting Indigenous peoples and other communities directly impacted by climate crisis" (30).

Today, brown is facilitator of Detroit Narrative Agency (DNA), "supporting Detroiters to shift the narratives of the city towards justice and liberation."[2] She has also been co-facilitator for the Detroit Food Justice Task Force, a coalition working together to improve food access and security, create jobs, and contribute to community sustainability. As well, they provide information and resources on food justice, environmental justice, and media literacy. In addition, she has been a facilitator for Detroit Future, a citywide movement to "build a more just, creative, and collaborative Detroit through broadband adoption activities"; the Detroit Digital Justice Coalition, which works on building "healthy digital ecologies" in which people learn to use the Internet to transform their communities; the Black Mesa Water Coalition, a group of young intertribal and interethnic people dedicated to addressing issues "of water depletion, natural resource exploitation, and health promotion within Navajo and Hopi communities"; and INCITE!, a "national activist organization of radical feminists of color" working to "end violence against women of color, trans & gender non-conforming people of color, and our communities through direct action, critical dialogue, and grassroots organizing."[3]

Brown builds on Butler's work to imagine the future of climate change in three especially illuminating ways. First, brown's approach is intersectional: she understands climate change cannot be isolated from other economic, social, scientific, and technological problems. In other words, confronting climate change necessarily involves confronting other inequalities. This clustering together of issues is apparent in the many different kinds of movements with which brown has worked. The insistence on not isolating climate change problems from larger economic, racial, and social problems and conflicts over colonialism is one of the biggest differences between mainstream environmental movements and movements that enjoy significant leadership from Indigenous people and people of color. Second, though brown is attuned to the particularities of place, she also thinks about climate change as a world problem and makes connections between particular places and the unequal ways climate change is affecting all of us. Her movements use a range of strategies, including appeals to law, nation-states, and international bodies as well as direct action to shape change in more basic ways not beholden to nation-states or wealthy stakeholders. Finally, brown works at intersections that centrally involve Indigenous people and people of color, thinking about complex relations and solidarities of many sorts, including around climate justice. Perhaps Butler's recurring stories of people making communities across differences in postapocalyptic conditions also helped inspire brown's work in this vein.

The post-2000 speculative texts I take up in this final chapter illuminate obstacles to nation-states solving climate change problems, point to direct action as a crucial method, and imagine other possible worlds rather than hoping that nation-states or captains of industry will save the day. This tension between broad understandings of climate justice as inseparable from decolonization, the redistribution of wealth, and the decentralization of power, insisted upon by movements led by Indigenous people and people of color, on the one hand, and narrower frameworks for imagining the future of climate change shaped by international bodies dependent on nation-states in thrall to the global fossil fuel economy, on the other, is one of the biggest obstacles we face in shaping the climate change disaster that both lies ahead of us and is happening right now.

IMAGINING THE FUTURE OF CLIMATE CHANGE AS A WORLD PROBLEM

After the 1992 Rio Earth Summit, significant efforts were made, especially through the UN, to present climate change as a world problem. One key flashpoint is the annual Conference of the Parties (COP) to the UN Framework Convention on Climate Change, which has taken place since 1995. From Copenhagen to Paris, Indigenous people and people of color have pushed international bodies and nation-states to use a climate justice prism to understand disaster as fundamentally shaped by race, settler colonialism, and class. They have also pushed for lower and binding carbon emissions limits and broadened visions of climate change. But despite significant achievements, such projects are severely limited by the unwillingness of states, especially where fossil fuel industry interests and lobbies dominate, to take decisive steps to lower carbon emissions and shift to renewable energy.

One of the most important international efforts was the 1997 Kyoto Protocol, which was adopted with more than 150 signatories. It included legally binding emissions targets for six major greenhouse gases for "developed country" parties. It also offered alternative ways of meeting emissions targets through market-based mechanisms such as cap and trade, which creates tradable pollution permits. The United States signed the Kyoto Protocol but Clinton never sent it to Congress for ratification. In 1997, the Senate had voted 95–0 expressing disapproval of any agreement that did not force developing countries to reduce emissions and which "would seriously harm the economy of the United States"—the latter an unspecified, ambiguous standard whose nebulousness was used to justify taking no action.[4] In 2001 George W. Bush withdrew U.S. support and today the United States remains the only signatory not to have ratified the protocol. The protocol was adopted by early 2005 but since, of the world's two largest polluters, the United States never ratified it and China never signed it, and Russia, Canada, and Japan abandoned it in 2011, its impact was limited. The protocol did, however, inspire low carbon laws in many places and according to the UN, emissions in signatory nations were 22.6 percent lower than 1990 levels in 2012, well beyond the 5 percent reduction commitment.[5] The protocol also established an adaptation fund and incentives for green investments in developing nations.

The question of how to address historical and contemporary global disparities in carbon emissions and wealth between industrialized and developing nations continues to be a sticking point. The 2009 Copenhagen conference failed to achieve binding agreements and many viewed this as the end of hopes for avoiding dangerous future climate change. During the summit's final hours, leaders from the United States, Brazil, China, Indonesia, India, and South Africa, excluding everyone else, agreed to the Copenhagen Accord, affirming the need to limit global temperature rise to 2°C, though no binding emissions limits were required, and gave $100 billion in climate aid to developing countries.[6] But just a few years later, at the 2012 Doha meeting, the United States, according to the Union of Concerned Scientists, "worked to prevent any explicit commitment by developed countries to ramp up their collective provision of climate finance" from current levels of $10 billion a year towards the $100 billion a year by 2020 pledged by Obama and other world leaders at Copenhagen.[7]

Other important recent flashpoints include the first Climate Justice Summit, held in 2000 in The Hague, which was imagined as a radical alternative to official talks and has been credited with bringing the idea of climate justice onto the global stage. It was organized by the Rising Tide Network, an "international, all-volunteer, grassroots network of groups and individuals who organize locally, promote community-based solutions to the climate crisis and take direct action to confront the root causes of climate change."[8] One participant, Ivonne Yanez from Oilwatch International in Ecuador, later recalled meeting Roberto Afanador Cobaria, a representative from the U'wa people from Colombia, who in 2002 succeeded in their effort to prevent an oil company from drilling on sacred land. Yanez says she remembers this summit as "the first time Oilwatch really started to argue that we must 'leave the oil in the soil' to address climate change." She lists several other successful climate justice battles that followed, including "the achievements of the Ogoni people against Shell, Costa Rica's decision to declare the country an oil-free zone, and the Bolivian government's decision to give greater rights to its indigenous peoples."[9]

At the Second Earth Summit the next year in Johannesburg, the International Climate Justice Network produced the Bali Principles of Climate Justice, which call for "an international movement of all peoples." Seeking to make climate change a human rights, environmental justice, and local communities issue, the coalition insisted that "climate change and its associated impacts are a global manifestation" of a "local chain of impacts," caused "primarily by industrialized nations and transnational corporations." They charged development banks, transnational corporations, and "Northern governments, particularly the United States" with compromising "the democratic nature of the United Nations as it attempts to address the problem," and claimed that "the perpetration of climate change violates the Universal Declaration on Human Rights and the United Nations

Convention on Genocide." They argued that climate change impacts "are disproportionately felt by small island states, women, youth, coastal peoples, local communities, indigenous peoples, fisherfolk, poor people and the elderly" and objected that "local communities, affected people, and indigenous peoples have been kept out of the global processes to address climate change." The coalition deplored "market-based mechanisms and technological 'fixes' currently being promoted by transnational corporations," such as geo-engineering, as "false solutions" that make matters worse; it also warned that climate change threatens "food sovereignty and the security of livelihoods of natural resource-based local economies" as well as "the health of communities around the world," especially the "vulnerable and marginalized, in particular children and elderly people." The coalition also emphasized that impacts will be particularly devastating in the global South and the "South within the North." Finally, the coalition called for "an international movement of all peoples for Climate Justice" based on twenty-seven core principles, including "the rights of indigenous peoples and affected communities to represent and speak for themselves"; the demand that "communities, particularly affected communities, play a leading role in national and international processes to address climate change"; condemnation of transnational corporations deepening the climate change disaster; a call for recognition of the "ecological debt" owed to "the rest of the world" as a result of privileged nations' "appropriation of the planet's capacity to absorb greenhouse gases"; and the demand for a "moratorium on all new fossil fuel exploration and exploitation."[10]

Demands for climate justice grew louder in the wake of Hurricane Katrina. In 2006, one year after Katrina, environmental justice scholars Manuel Pastor, Robert D. Bullard, James K.

Boyce, Alice Fothergill, Rachel Morello-Frosch, and Beverly Wright published *In the Wake of the Storm: Environment, Disaster, and Race after Katrina,* explaining in their introduction that while Katrina "shattered the illusions many Americans have about disaster," including that they are "an equal opportunity affair" rather than being "heavily affected by income and race," environmental justice scholars and activists were not surprised at all. On the contrary, over the last twenty years they had compiled a vast amount of research showing that "disparities in environmental conditions were a worrisome norm in many parts of the United States." When the hurricane landed, they argued, "existing inequalities and the history of discrimination in the American South played out in tragic yet predictable ways," leaving the most vulnerable inadequately protected.[11]

That same year, the Inuit filed an appeal with the Inter-American Commission on Human Rights "seeking relief from violations of the human rights of Inuit resulting from global warming caused by greenhouse gas emissions" from the United States.[12] Led by Sheila Watt-Cloutier, elected chair of the Inuit Circumpolar Council (ICC), which represents 160,000 Inuit from Alaska, Canada, Greenland, and Russia, the Inuit named the United States as the biggest emitter of greenhouse gases that refused to set emission targets and backed off from its Kyoto Protocol commitments. In 2006, the commission declined to rule, saying there was "insufficient evidence of harm."[13] Even so, the petition called a world audience's attention to the dramatic ways global warming was affecting Indigenous people in the Arctic.

The 2007 adoption of the UN Declaration on the Rights of Indigenous Peoples was an historic event with important implications for environmental and climate justice struggles. Article 29 of the declaration states that Indigenous peoples have the "right to the conservation and protection of the environment and the productive capacity of their lands or territories and resources" and recommends states establish assistance programs for Indigenous peoples and "ensure that no storage or disposal of hazardous materials" take place in their lands or without their consent. Article 32 declares states must cooperate "in good faith with Indigenous people" and get their consent before approving "any project" affecting their lands and territories, "particularly in connection" with the development of "mineral, water or other resources." Even though the United States, Canada, New Zealand, and Australia voted against it, the declaration passed by a majority vote.[14]

A few years later, at the 2010 COP16 meeting in Cancún, the ICC released a document identifying itself as the "international voice of Inuit" and "calling upon global leaders at the December UN Climate Change Conference of the Parties to listen to this Inuit voice." The Inuit "Call to Global Leaders" urged immediate action on climate change and recommended setting binding greenhouse gas emissions reduction targets to limit global temperature rise to 1.5 degrees Celsius, providing financial support to help Inuit adapt, placing decision-making and funding at the lowest possible levels (communities instead of states), recognizing the role of the Arctic in sustaining global climate systems, designating avoiding further climate change in the Arctic as a key progress benchmark, integrating traditional and local knowledge into the IPCC reporting process, and providing support for local green energy. The ICC also insisted that "any adaptation framework adopted by the global community should recognize the responsibility of wealthy countries towards communities within their borders that are the most vulnerable to climate change impacts, including indigenous peoples." It should recognize as well "the vulnerability of peoples, rather than nations, to the impacts of climate change to highlight the differential impact at the regional and local, rather than national level."[15]

The 2015 Paris meeting, COP21, resulted in a new global agreement among 195 countries to limit global temperature increase to 2 degrees Celsius and, if possible, to 1.5 degrees. Greater increases, the participants warned, would catastrophically affect the planet, leading to dramatic sea level rise, destructive floods, storms, droughts, and food and water shortages, while even a 1.5 degree warming might mean the end for low-lying island nations. At this meeting, several Inuit leaders participated, including Natan Obed, president of Canada's national Inuit organization, who was part of Canada's official delegation. Inuit youth also attended, including Maatalii Okalik, president of Canada's National Inuit Youth Council. "Inuit continue to be the human barometer of climate change," she claimed during the conference, adding that Inuit had been telling "the international community for years that climate change is happening at a rapid pace."[16] In its 2015–16 report, the Inuit Circumpolar Council said their global advocacy work had resulted in "the preamble of the Paris Agreement" stating that "the rights of Indigenous Peoples should be respected" and that "climate adaptation should be based on the best available science," including "the knowledge of Indigenous Peoples."[17]

Indigenous people and people of color have often led efforts to create solutions to climate change as a world problem, despite the impediments created by corporations and large nation-state greenhouse gas emitters. The unwillingness of states and the fossil fuel industry to change and the current conjuncture, in which Trump's administration will make matters dramatically worse, however, has many activists, artists, and writers looking elsewhere for truly transformative change.

SOCIAL MOVEMENTS, DIRECT ACTION, AND BLACK FEMINIST FUTURISMS

In conclusion, I return to the luminous work of adrienne maree brown in imagining the future of climate change, making different worlds through direct action and social movement-building, and creating transformative change through visionary speculative fiction. Brown first became visible as a climate justice, youth, and social movement organizer in the early 2000s as cofounder and director of the League of Young/Pissed Off Voters, a national organization focusing on electoral reform and campus organizing. She also coedited the youth organizing collection *How to Get Stupid White Men out of Office.* And in 2006, she worked with the Arctic Indigenous Youth Alliance, writing in her blog that "the land is melting out from under them while major corporations attempt to build oil pipelines through their sacred spaces" and that "these young people are banding together from different super rural locations to try to leverage power for and with their people, so it's quite exciting."[18]

She draws on this work with youth in her essay, "The Green Generation," for Bill McKibben's 2007 *The Global Warming Reader: A Century of Writing about Climate Change.* Brown focuses on "today's generation" of sixteen- to thirty-five-year-olds, arguing they will uncover answers to many tough questions: "How does a planet move from consumption to sustainability?" And "will we make the connection between global warming and economic disparity on a large enough scale?" How can we make that connection when "the only thing incremental about climate change is our American response to it"?[19] Sounding a hopeful note, brown promises that "this generation is teeming with activists and organizers who will make the big visionary connection and take the drastic strategic actions on behalf of our survival" (260). Calling attention to youth who experience climate change directly, she emphasized how "in the Arctic and the Gulf Coast, youth have seen drastic changes in their land, homes, infrastructure, weather." In "urban industrial and postindustrial places like Detroit, NY, Chi, and Pitt," she clarifies, "youth experience the physical build-up of pollution in their bodies." Others "experience it less directly—targeted heavily by military recruiters desperate to beef up forces for wars to secure our prolonged reliance on outdated resources" (260). She interviews a wide variety of youth and records their knowledge about how climate change is transforming their worlds. "In each space," she concludes, "there is more action than I've ever seen before by young people from impacted communities" who "are looking for people to move the front line of their sustainable visions forward against all odds" (267).

Teaching direct action is central to the creative and future-shaping work brown does with youth and other movements. In a 2009 interview in the "Voices of Climate Change" series, brown explains she came to Detroit in 2006 to do organizational development and for direct action trainings with Detroit Summer, a project cofounded by Grace Lee Boggs and Michelle Brown. "More than forty years ago, in the wake of urban rebellions," Boggs remembered, "Dr. King said that young people 'in our dying cities' need direct action projects that transform themselves and their surroundings at the same time."[20] Inspired by this example, in 1992 Detroit Summer, a multiracial, intergenerational collective, worked to transform communities through youth leadership, creativity, and collective action. "We wanted to engage young people in community-building activities: planting community gardens, recycling waste, organizing neighborhood arts and health festivals, rehabbing houses, painting public murals," Boggs recalled. "Encouraging them to exercise their Soul Power would get their cognitive juices flowing." Thus "learning would come from practice, which has always been the best way to learn."[21]

Direct action was important for brown when she was executive director of the Ruckus Society from 2006 to 2010 and sat on their board through 2012. The Ruckus Society, formed in Oregon in 1995, provides environmental, human rights, and social justice movement organizers with tools, training, and support through strategic use of creative, nonviolent direct action. Created in response to an anti-environment, pro-logging bill signed by President Clinton that galvanized intense resistance, they train and support communities ranging from "frontline climate change activists from the Arctic North to the Gulf Coast, to Latina garment workers in LA, day laborers in the SF Bay Area, steelworkers in Indiana, student organizers in NYC, Hip Hop artists from both coasts, conscientious objectors in the Heartland, and Indigenous organizers across North America."[22] Brown explains why direct action is so important to the visionary futurisms she co-creates with communities: "Direct action is where escalation happens, where people can play an active role in advancing a negotiation, where we see and feel each other's solidarity." She reminds us that direct action was key to civil rights movements, when "we first saw images of blacks and whites at lunch counters together in the south." She argues that today, "guerilla gardens," like those Octavia E. Butler imagined as part of a climate change future in *Parable of the Sower,* are "a way to show that we know how to live more sustainably and we will push our leaders to catch up with us." For brown, direct action means living and cocreating the world you want to see:

"Our actions have to be towards the world we want," she insists. "We need to be guerilla gardening and turning people's heat and water on. We need to be the guerillas putting up solar panels in the hood." Rather than trying to "fit into someone's assembly line and make things for the class above us," brown understands direct action as indispensable to liberation, with the latter defined as freedom "to work for our own communities, to thrive, to be in symbiotic relationships based on our needs and our dreams."[23]

In her movement-building work brown emphasizes the importance of intersecting social justice struggles. She argues it is impossible to confront Hurricane Katrina without accounting for "how we look at race in this country" and that movements for "climate justice" happening across the United States at the time were "all about race." If you look at "who's impacted by climate injustice, who's impacted by the war, who's impacted by our economic policies, and then who's being disenfranchised at election time, it's the same lines," she suggests. "From our work within the Indigenous People's Power Project, and now learning with Movement Generation, it's clear that those issues naturally intersect." Although people often "try isolated organizing," brown is convinced "we have to start seeing isolated issues in the larger context of ecological justice for all." Imagining the future of climate change in ways that creatively swerve outside the frameworks established by nation-states, brown advises "we need movements that aren't centered on what the powers that be can grant us, but rather on what we can build and practice together." In terms of sustainability, she argues, "I don't believe we can have a green future or any future, unless we understand that we have to change the power dynamics based on race, class, and gender."[24]

As brown's organizing suggests, connecting Indigenous peoples' struggles over climate change and the environment to those of people of color in the United States is one of her major contributions. Reflecting on this collective project in *Emergent Strategy,* brown observes, "Every time I have worked with Indigenous communities that have been able to sustain their cultural practices through the onslaught of colonialism and imperialism, as I listen, I hear emergent strategy—being in right relationship with the natural world, learning from the ways change and resilience happen throughout this entire interconnected complex system" (26). One of the biggest problems today, according to brown, is how non-Indigenous people "learn to disrespect Indigenous and direct ties to land" (47). She states that her work with the Indigenous People's Power Project (IP3) was transformative for her organizing and vision of the world. The collective's aim was "to build a body of Indigenous organizers who became action experts within their own communities" (61). In the process, she and Ruckus "learned a lot about breaking down the walls between different issue areas," since "Indigenous communities present a clear case of economic and environmental hardship, with residents highly recruited for the military, dealing with high levels of drug and alcohol dependence and a high rate of suicide" (62). These intersections clarified for her that "one piece of analysis" won't "serve the big picture" and that "for successful movements, we need to develop strong, action-oriented communities that understand that their analysis and work cannot be limited to one struggle" (62).

Brown also testifies that "the first time I heard about protocols was while working with Indigenous communities" and that she was "moved by the clarity of the protocols—in each community there are ways to honor and respect the culture, the elders, the leadership, the history, and the power dynamics." Brown went on to "experience protocols being practiced in intersectional ally work (226). She claims that Ruckus's work with Indigenous groups made the "small organization" grow "from a kickass, majority white, male-led environmental issue-centered network into a kickass, female-led, multicultural, justice- AND environment-centered network" (65).

Since 2010, brown has facilitated the connection of groups and issues around climate justice, social change, movement-building, and speculative fiction in a number of ways. In 2013, she received a Detroit Knight Arts Challenge Award to run a series of Octavia Butler-based science fiction writing workshops. She also organized Octavia Butler and Emergent Strategy reading groups for people hoping to use Butler's work to imagine political and strategic solutions to problems. As well, she has collaborated with Ayana Jamieson, the founder of the Octavia E. Butler Legacy Network, a digital hub that connects artists, activists, scholars, and others who extend Butler's work in the direction of creating and connecting communities. She was central to the conference Jamieson and I coorganized at the University of California, San Diego in June 2016, sponsored by the Arthur C. Clarke Center for Human Imagination and the Clarion Workshop. At that conference, "Shaping Change: Remembering Octavia E. Butler through Archives, Art, and World-making," this network was powerfully embodied by filmmakers, students, artists, archivists, activists, scholars, and writers inspired by Butler and doing work in her memory over three days filled with brilliant presentations and stimulating collective conversations.

In an interview with Moya Bailey in a 2013 special issue on "Feminist Science Fiction" in *ADA: A Journal of Gender, New*

Media, and Technology, brown talks about using "Butler and speculative fiction to do this work" and of her growing "suspicion that the realm of science and speculative fiction could be a great place to intentionally practice the futures we long for."[25] Bailey explains how brown "brings communities together through the thread of Octavia Butler's writing in collaborative sessions that emerge around the curated content of her *Octavia Butler Strategic Reader.*" In 2010, "brown facilitated the first session that created the reader at the Allied Media Conference in Detroit," collectively producing a crowd-sourced document edited by brown and Alexis Pauline Gumbs and "made available online for free to anyone interested in working with the texts." Bailey remembers that brown "helped hold the space, but required the active participation of all the people in the room," co-creating "a nonhierarchical, interdependent community as we experienced multiple forms of knowledge creation and expansion." This creative organizing activity has subsequently morphed into brown's emergent strategy sessions as well as the collective project of *Octavia's Brood: Science Fiction Stories from Social Justice Movements.*

In her "Outro" to *Octavia's Brood,* brown writes of "the abundance of imagination we in the social justice realm hold, and must cultivate" as "seeds for the type of justice we want and need."[26] If we "want to bring new worlds into existence," she insists, "we need to challenge the narratives that uphold current power dynamics and patterns." Tools that help "bring the work off the page and into our lives" include visionary fiction, which "explores current social issues through the lens of sci-fi; is conscious of identities and intersecting identities; centers those who have been marginalized; is aware of power inequalities; is realistic and hard but hopeful; shows change from the bottom up rather than the top down; highlights that change is collective;

and is not neutral—its purpose is social change and societal transformation" (279). Another tool is emergent strategy, which is "intentional, interdependent and relational, adaptive, resilient because it is decentralized, fractal, uses transformative justice, and creates more possibilities" (280). A third tool are science fiction or visionary fiction writing workshops designed to produce ideas collectively, identify issues important to local communities, and build "a world in which to explore the issue and possible solutions" (281).

I first met brown briefly at the American Studies Association's 2014 conference in Los Angeles, where she, Jamieson, Imarisha, and Bailey led a session devoted to "bringing Octavia strategies for liberation to the academy," as brown would later put it in her blog.[27] Then in 2015, she attended the Clarion Science Fiction and Fantasy Writers Workshop, the oldest and most prestigious workshop of its kind in the world, which I have directed since 2010. I was excited she was joining the class because of her stellar submission stories, including "The River," which appears in *Octavia's Brood.* There was already a buzz about *Octavia's Brood* and the anthology made quite a few reading lists of exciting new books to read that year.

In the anthology, brown and Oakland-based journalist Dani McClain, who covers reproductive rights, sexuality, and education, imagine near-future worlds dramatically reshaped by changes in nature and climate. In "Homing Instinct," McClain explores dystopian possibilities of state power being extended in unjust ways in a future where climate change has drastically worsened and tries to imagine how her protagonist might get free. On the other hand, in "The River," brown conceives of the Detroit River as having a kind of personhood, as having a kinship with the people of Detroit, so much so that it rises up tsunami-style to wipe out disaster capitalists, thereby allowing those who remain to make another world. Both stories show the importance of direct action and speculative imaginings as critical resources at a time when nation-states are failing to make necessary changes in the face of climate change disaster.

In "Homing Instinct," McClain extrapolates from her present to imagine how an executive order might radically limit people's freedom of movement when the United States finally confronts climate change imminent disaster after The Dithering in our own times. As her Oakland-based protagonist lets Executive Order 2735's meaning sink in, she gets "greedy for things that likely won't be around much longer," like "70-degree days in the middle of February," a familiar sensation in our greenhouse world.[28] The executive order, issued by a female president without a Congressional vote, gives everyone ninety days to "reposition themselves": Operation HOMES: (Honoring Our Most Enduring Settings) mandates that people register their locations in a database and travel only rarely using mile allotments, thereby "making mobility a luxury item" (241). Resisters who refuse to register are shipped off to upstate or out-of-state penitentiaries. The new law caps "all oil dependent travel at twenty miles per month." Getting people away from the coasts, "the places the oceans reclaimed for themselves more and more each year," is a priority, so sky miles are added to allotments for "people who agree to move away from the Gulf coast and parts of California and the eastern seaboard" (242).

Choosing a home is not difficult for the protagonist's bestie, Paloma, who argues that issuing an executive order is wise because otherwise climate change deniers would prevent any action from being taken. She's grateful that the president is forcing people to "figure out where you want to be and get there."

The protagonist momentarily thinks to herself that Paloma should be featured in PSAs for the new executive order, along with "quick takes from the footage of the latest disasters no one even bothered calling 'natural' anymore": "giant waves whipping into the now-nonexistent Outer Banks of North Carolina"; the "Miami beach exodus"; and the "now iconic images of people swimming in the streets of New Orleans." The new executive order plays off of "people's fear" and "memories of disasters," the narrator realizes, yet at the same time she concedes "the relationship between water and land, between humans and the weather, had changed dramatically" and "yes it was long overdue for a political leader to demand that people stop living in the fantasy of the infinite" (240).

But after talking on the phone to her faraway mother, who encouraged her wanderings and will accept her decision about where home is, the protagonist decides even though "the law was right to urge people to think about where the land could actually sustain them," it was also "an immoral law predicated on an outdated belief in stability." Rejecting the "fantasy that immobility" will "bring security" as well as the idea of "buffering oneself from the chaos and destruction that had come to define the times," the protagonist decides to choose "path as a possibility" and remembers "Fred Korematsu, Assata Shakur" and "others who had escaped, run, resisted" (246).[29] Although the protagonist makes fun of Paloma for taking seriously an Internet message from Native people somewhere in Arizona asking the "pointed questions: Where are you living? What are you doing? What are your relationships? Where is your water?" she ultimately discovers the message has lodged in her mind, "finding a place to settle amidst the cynicism, fear, and doubt" (245). At the end, she resolves to resist the Executive Order and run, planning to "push off towards the middle of the river" and wondering who she will see "there in the middle of the torrent" (247).

If the river is the pathway of escape from state power and the means of making alternative community in "Homing Instinct," in brown's "The River" it is alive, haunted by history, and the agent of transformative change enabling a radical reimagining of Detroit's future in opposition to neoliberal visions. Brown's protagonist is a "water woman" who "felt rooted in the weather," having learned "life lessons" from her grandfather.[30] At the beginning, she reflects on how when out on the waves she keeps feeling "something in the river" growing, haunting "the island between the city and the border" (23). The island is Belle Isle, the largest city-owned island park in the nation, designed by Frederick Law Olmsted, who also designed Manhattan's Central Park. Brown describes it as an "overgrown island," housing the "ruins of a zoo, an aquarium, a conservatory, and the old yacht club, down the way" from "the abandoned, squatted towers of the renaissance center, the tallest ode to economic crisis in the world" (23–24). The Renaissance Center, now General Motors' headquarters, is a group of seven interconnected skyscrapers located on the Detroit waterfront. By imagining the RenCenter abandoned by capitalists and taken over by squatters, brown situates the story in a near future that is our world, but worse. The Detroit River was a site of dumping and other kinds of industrial pollution for decades, though cleanup efforts to dredge pollution out of the river have helped in the last twenty years and some animal species are now returning. Perhaps the water woman is thinking of this when she reflects "it made a sort of sense that something would grow down there," since "nuf things went in for something to have created itself down there." She also remembers her grandfather telling her even though they had to "fight for any inches we get" in this country, "the water has always helped us get free one way or another" (23). The Detroit River was an important conduit on the Underground Railroad, the pathway to freedom in Canada during slavery times, and the water woman remembers that history, loving "to anchor near the Underground Railroad memorial and imagine runaway slaves standing on one bank and how good the water must have felt" (24). When she talks to her friend about her feelings, he says that in "certain parts, it's like an ancestral burial ground" and "a holy vortex of energy," but she internally disagrees, thinking to herself that the river feels "alive and other" (25) rather than dead.

The near-future Detroit that brown imagines, a "broke city" that turned off the streetlights, is in the grips of disaster capitalism. "Folks born and raised here" can't make a living or get investors, but "she heard entrepreneurs on the news talk of Detroit as this exciting new blank canvas," which makes her wonder "if the new folks just couldn't see all the people there, the signs everywhere that there was history and there was a people still living all over that canvas" (26). The Detroit River soon reminds them of this, however, as it rises up and swallows newcomers "drawn by the promise of empty land and easy business, the opportunity available among the ruins of other people's lives." In the climax, the water sweeps away "the third consecutive white mayor of the great black city," who lives in a "mansion with a massive yard and covered dock on the river" overlooking Belle Isle and Canada. Born "in Grand Rapids, raised in New York, and appointed by the governor, he'd entered office with economic promises on his lips, as usual, but so far he'd just closed a few schools and added a third incinerator tower to expand Detroit's growing industry as leading trash processor of North America" (28). Out in her boat the night it happens, the water woman is "close enough to see it" when the waves start to "swell erratically," continuing to "rise and role," until finally she witnesses "one solitary and massive wave, a sickly bright green whip up out of the blue river, headed toward the mayor's back" (29). The wave, no wider than the house, takes the mayor, his wife, and a few others with it and then recedes as the other guests scramble to safety.

After the tsunami, the public panics and the island is closed with iron blockades, but new possibilities emerge in the wake of disaster. The "newly sworn-in mayor" calls the tsunami "an opportunity, wrapped in a crisis, to take the city back." The water woman watched "investors and pioneers pack up, looking for fertile new ground" and "she noticed who stayed, and it was the same people who had always been there, a little unsure of the future maybe, but too deeply rooted to move anywhere quickly," wondering why the river never touched them and grateful that they "got" their "city back" (31). The story ends with the water woman out in her boat, "searching inside the river, which was her most constant companion" for clues. Every now and then, she catches sight of "something swallowed, caught, held down, so the city could survive, something that never died, something alive" (31).

Brown's Detroit River is alive and on the side of people with history in the city. When it rises up tsunami-like, it destroys disaster capitalism's characteristic forms of development and political governance, allowing locally rooted people to take back their city. Instead of a mayor raised far from Detroit and appointed by the governor, the city leader is now a local man involved in gardening movements in which brown herself participated, connected to direct action community projects created by Grace Lee Boggs and others. Early on, the water woman remembers how her mama taught her important things like how to love Detroit and "that gardening in their backyard was not a hobby but a strategy" (24). Gardening is only one of many local strategies the people of Detroit have used to cultivate communities at a distance from state power.[31] In these ways, brown reminds her readers how the key to surviving disaster is making movements that center on what people can create together rather than what powerful nation-states and corporations are willing to give.

In her blog, reflecting on the Army Corp of Engineers' decision to stop DAPL construction pending an environmental impact report, brown reproved outsiders who wanted to rush to judgment that this was not a true victory because it would be overturned once Trump took office. Calling attention to Facebook live video, which she had reposted and recommended watching, she cites the testimony of Tokata Iron Eyes, a thirteen-year-old member of the Standing Rock Sioux tribe, who said "I feel like I have my future back!"[32] They "don't say these things because they lack context or information or misunderstand the patterns of this country and need non-native people to educate them," brown emphasizes, "they say these things with lifelong experiences of being in this battle for the planet, against nations." The "victories are few" but they "nourish us" and "help us to understand the potential of intersectional peoples' power," so "we have to protect the time and space needed to celebrate." Connecting many of the most important struggles for climate justice across space and time through her intersectional movement-building and speculative fiction, brown helps communities "flex their collective imagination muscle" and envision "creative direct actions." In the face of the failures of nation-states and international bodies to set binding carbon emissions limits that might save the planet, these kinds of networked local strategies, direct actions, and collective envisionings of the future may well be our best hope in imagining other worlds in the wake of the climate change disaster that is now upon us.

ACKNOWLEDGMENTS

This book would not have been possible without the help and inspiration of large numbers of people. I wish I could thank them all here but there is not enough space. What follows is a truncated accounting for some of the manifold forms of symbiosis I have enjoyed and benefited from while writing this book.

Becoming faculty director of the Clarion Science Fiction and Fantasy Writers Workshop at the University of California, San Diego changed my life. I am grateful to Kim Stanley Robinson and the Clarion Foundation for bringing Clarion here. Over the last seven years, Stan has generously visited my classes, spent serious time with my undergraduate and graduate students, given many lectures for free, and worked tirelessly to keep Clarion going. Karen Joy Fowler, Clarion's president, has also visited my ethnic studies and literature classes, donated her time to speak to huge numbers of undergraduates, and is an inspiration and a role model for me—she's one of the best Clarion teachers with whom I have had the privilege of sharing a classroom. I owe thanks to former Clarion coordinator Laura Martin for being such a terrific partner in running Clarion and to new coordinator Patrick Coleman for his indispensable help. I also want to thank Sheldon Brown, director of the Arthur C. Clarke Center for Human Imagination, for all his support in providing a home for Clarion at UCSD.

Most of all, I want to thank all the writers I have met at Clarion since 2010, including adrienne maree brown, whose work inspired chapter 3. The Clarion world has nourished this book and *Imagining the Future of Climate Change* would not be possible without it.

I owe the deepest thanks to Ayana Jamieson, the founder of the Octavia E. Butler Legacy Network, whose own work and our ongoing collaborations and friendship have been crucial to my thinking about this book and many of the other intellectual projects about which I care most. Organizing the "Shaping Change: Remembering Octavia E. Butler through Art, Activism, and World-Making Conference" at UCSD in June 2016 with Ayana was definitely a highlight for me. All of the people who participated in that life-changing conference shaped this book. I also want to thank the Arthur C. Clarke Center for Human Imagination and Clarion (again!) for making the conference possible with material support.

I can only gesture toward the world of writers, artists, filmmakers, and producers of speculative fiction who have impacted this book and to whom I owe thanks. Among those who have taught at Clarion, Nalo Hopkinson and Ted Chiang have been especially generous interlocutors in ongoing conversations and I know how lucky I am to be able to say that. Alex Rivera, director of the genius film *Sleep Dealer*, generously donated his time for a series of events at UCSD and also met with my 2015 Clarion students, inspiring both them and me. Alexis Lothian and Aimee Bahng are two colleagues with whom I am especially fortunate to be in conversation: thinking together for conference panels and projects over the last couple years has been a real joy, and their work has pushed mine in many good directions.

I want to thank Rosaura Sanchez and Beatrice Pita for continuing to make a science fiction studies cluster at UCSD possible through their own brilliant contributions to the field and their work with our students. I also want to thank my colleagues and students in literature and ethnic studies at UCSD, especially the students in my Winter 2017 ethnic studies seminar on social movements and culture, for stimulating conversations as I was finishing this book.

I am glad to be in this University of California Press series, which already features two excellent books on social movements by Scott

Kurashige and Rod Ferguson. I am grateful for Lisa Duggan's editorial expertise and Niels Hooper's steady guidance throughout the whole process. I also greatly appreciated Caroline Knapp's compassionate and skillful copyediting at the end.

Conversations, hatching plans, and sharing ideas about teaching science fiction with Chris Cunningham are all at the heart of this book. I also need to thank my dad, Jim Streeby, who inspires me with his love of mentoring people inside and outside the bio-family and who talks to me on the phone almost every day about my research, writing, and Clarion. It is wonderful that I can share so many things with my great aunt Sheila, also a generous mentor to many, and I appreciate her care and concern for me as well as for the future of the planet and all its critters. My ongoing conversations with my youngest brother, Patrick, also shaped this book. We, too, share a great deal and I am so glad he is there. As well, while writing this book, I thought a lot about the future world I hope my nieflings, Kathryn, Brittany, Evan, Wyatt, Avery, Reed, Aurora, Jordan, and Megan, will see and help to make.

Every day, I am grateful and lucky beyond measure for Curtis Marez's brave and brilliant work, nourishing the creativity of myself and many other people, and for our life in the desert and the green world.

NOTES

INTRODUCTION

1. See also Gerry Canavan, "If the Engine Ever Stops, We'd All Die": *Snowpiercer* and Necrofuturism," *Paradoxa* 26 (2014): 41–66.

2. Clive Hamilton, *Earthmasters: The Dawn of the Age of Climate Engineering (*New Haven: Yale University Press, 2013), 1. Hereafter page numbers appear parenthetically in text. See also Naomi Klein, *This Changes Everything: Capitalism vs. the Climate* (New York: Simon and Schuster, 2014), 57; James Roger Fleming, *Fixing the Sky: The Checkered History of Weather and Climate Control* (New York: Columbia, 2012); and Mike Hulme, *Why We Disagree about Climate Change: Understanding Controversy, Inaction and Opportunity* (Cambridge: Cambridge University Press, 2009).

3. Credits and production notes, *Snowpiercer* (South Korea), Radius TWC, 2014 , Margaret Herrick Library, Beverly Hills, California.

4. Brian Forno, "Snowpiercer: Director Bong Joon-ho on Cute Dictators," *Crave,* June 23, 2014, www.craveonline.com/site/710611-snowpiercer-bong-joon-ho-on-cute-dictators#A2kAVg040iL48E28.99, and Charlie Jane Anders, "Bong Joon-ho Explains Why *Snowpiercer*'s Violence Is So 'Explosive,'" June 25, 2014, *Io9,* http://io9.gizmodo.com/how-bong-joon-ho-turned-snowpiercer-into-your-worst-dys-1596079364.

5. Forno, "Snowpiercer."

6. Jean Noh, "Bong Joon Ho, Snowpiercer," *Screen International* (Aug.-Sept. 2013): 12.

7. On cli-fi and climate change literature, see Adeline Johns-Putra, "Climate Change in Literature and Literary Studies: From Cli-Fi, Climate Change Theater and Ecopoetry to Ecocriticism and Climate Change Criticism," *WIREs Climate Change* 7, no. 2 (March/April 2016): 266–82; E. Ann Kaplan, *Climate Trauma: Foreseeing the Future in Dystopian Film and Fiction* (Rutgers, NJ: Rutgers University Press, 2015); Richard Kerridge, "Ecological Approaches to Literary Form and Genre: Urgency, Depth, Provisionality, Temporality," in *Oxford Handbook of Ecocriticism,* edited by Greg Garrard, 361–75 (New York: Oxford University Press, 2014); and Sylvia Mayer and Alexa Weik von Mossner, *The Anticipation of Catastrophe: Environmental Risk in North American Literature and Culture* (Heidelberg: Universitätsverlag Winter, 2014).

8. Amitav Ghosh, *The Great Derangement: Climate Change and the Unthinkable* (Chicago: University of Chicago Press, 2016), 72.

9. Patrick Wolfe, "Settler Colonialism and the Elimination of the Native," *Journal of Genocide Research* 8, no. 4 (2006): 387–409; J. Kēhaulani Kauanui and Patrick Wolfe, "Settler Colonialism Then and Now," *Politica and Societa* 2 (2012): 235–58.

10. Mark Maslin, *Climate Change: A Very Short Introduction* (Oxford: Oxford University Press 2014), 2.

11. David Archer and Stefan Rahmstorf, *The Climate Crisis: An Introductory Guide to Climate Change* (Cambridge: Cambridge University Press, 2010), 8.

12. Gilbert Plass, "Carbon Dioxide and Climate," *Scientific American* 201, no. 1 (July 1959): 41–47.

13. Bill McKibben and Dave Keeling, "The Keeling Curve," in *The Global Warming Reader: A Century of Writing about Climate Change,* edited by Bill McKibben (New York: Penguin, 2012), 44–45.

14. "Restoring the Quality of Our Environment," *Report of the Environmental Pollution Panel President's Science Advisory Committee,* November 1965. Hereafter citations appear parenthetically in text.

15. Wallace Broecker, "Climatic Change: Are We on the Brink of a Pronounced Global Warming?" *Science,* August 8, 1975: 461.

16. Jule G. Charney, Akio Arakawa, D. James Baker, Bert Bolin, Robert E. Dickinson, Richard M. Goody, Cecil E. Leith, Henry M. Stommel, and Carl I. Wunsch, *Carbon Dioxide and Climate: A Scientific Assessment, National Research Council, Ad Hoc Study Group on Carbon Dioxide and Climate* (Washington, DC: National Academy Press, 1979).

17. National Research Council, *Changing Climate: Report of the Carbon Dioxide Assessment Committee* (Washington, DC: National Academy Press, 1983), 1–2.

18. Octavia Butler once remarked that the "Global Climate Coalition " was "an anti-change group that could be called the Global Fossil Fuels Coalition." OEB 1921, Octavia E. Butler Papers, The Huntington Library, San Marino, California.

19. Andrew Revkin, "Industry Ignored Its Scientists on Climate," *New York Times,* April 29, 2009, A1.

20. Quoted in Revkin, "Industry Ignored."

21. Rachel Carson, *Silent Spring* (New York: Houghton Mifflin, 2002; original ed. 1962). Hereafter page numbers appear parenthetically in text.

22. Coll. 455, Box 12, Folder 18, James Tiptree Papers, Knight Library, University of Oregon.

23. Linda Lear, *Rachel Carson: Witness for Nature* (New York: Henry Holt, 1997), 408.

24. Jack Gould, "TV: Controversy over Pesticide Danger Weighed," *New York Times,* April 4, 1963, 95.

25. "The Desolate Year," *Monsanto Magazine,* October 1962: 4–9.

26. Paola Bacigalupi, "Foreword," in *Loosed Upon the World: The Saga Anthology of Climate Fiction,* edited by John Joseph Adams, xiv (New York: Saga, 2015).

27. Samuel Delany, "Some Presumptuous Approaches to Science Fiction," *Starboard Wine: More Notes on the Language of Science Fiction* (Wesleyan, CT: Wesleyan University Press, 2009), 26. In the 1991 notebooks Octavia Butler kept while writing *Parable of the Sower,* she wrote that Delany, who was her teacher at the Clarion Science Fiction and Fantasy Writers Workshop, said that "the purpose of abstraction is to

clarify, emphasize, or present alternate aspects of reality," making us "take a good look at the real by some deliberate distortion." See OEB 3248, commonplace book (large), Octavia E. Butler Papers, The Huntington Library, San Marino, California.

28. Fredric Jameson, *Archaeologies of the Future: The Desire Called Utopia* (New York: Verso, 2005).

29. Robert Heinlein, Letter, March 4, 1949, in *Grumbles from the Grave,* edited by Virginia Heinlein, 49 (New York: Del Rey, 1989).

30. Cecilia Mancusa, "Speculative or Science Fiction?" *The Guardian,* August 10, 2015; Margaret Atwood, "Writing Utopia," *Moving Targets: Writing with Intent, 1982–2004* (Toronto: Anansi, 2004), 92.

31. Ursula K. Le Guin, "The Year of the Flood by Margaret Atwood," *The Guardian,* August 29, 2009, www.theguardian.com/books/2009/aug/29/margaret-atwood-year-of-flood.

32. Hugo Gernsback, "A New Sort of Magazine," *Amazing Stories,* April 1926.

33. OEB 3085, *Parable* talk: early version: speech: notecard, Octavia E. Butler Papers, The Huntington Library, San Marino, California.

34. OEB 3090, Science Fiction: notecards, Octavia E. Butler Papers, The Huntington Library, San Marino, California.

35. OEB 116, "Black to the Future," Octavia E. Butler Papers, The Huntington Library, San Marino, California.

36. Work in the environmental humanities both within and outside of American studies has also been crucial. I cannot do justice to this large body of important scholarship here but one great place to start in thinking about climate change in humanities disciplines and in culture is Stephen Siperstein, Shane Hall, and Stephanie LeMenager, eds., *Teaching Climate Change in the Humanities* (New York: Routledge, 2017).

37. Raffaella Baccolini and Tom Moylan, eds., *Dark Horizons: Science Fiction and the Dystopian Imagination* (London: Routledge, 2003).

38. Butler's theorization of slow violence anticipates by two decades Rob Nixon's award-winning book *Slow Violence and the Environmentalism of the Poor* (Cambridge, MA: Harvard University Press, 2013), in which he conceptualized climate change as a kind of slow violence that is "typically not viewed as violence at all" because it "occurs grad-

ually and out of sight, a violence of delayed destruction that is dispersed across time and space" (2).

39. OEB 3193, commonplace books (medium), Octavia E. Butler Papers, The Huntington Library, San Marino, California.

CHAPTER ONE. #NODAPL

1. Robin W. Kimmerer, Melissa K. Nelson, Kyle P. Whyte, and Rosalyn LaPier, "Let Our Indigenous Voices Be Heard," www.esf.edu/indigenous-science-letter/Indigenous_Science_Declaration.pdf. See also Jace Weaver, ed., *Defending Mother Earth: Native American Perspectives on Environmental Justice* (Maryknoll, NY: Orbis Books, 1996).

2. Terri Hansen, "Marching for Indigenous Science," *Earth Island Journal,* April 21, 2017, www.earthisland.org/journal/index.php/elist/eListRead/marching_for_indigenous_science/.

3. Naomi Klein, *This Changes Everything: Capitalism vs. the Climate* (New York: Simon and Schuster, 2014), 328.

4. "Sacred Stone Camp," http://sacredstonecamp.org.

5. Phil McKenna, "Dakota Pipeline Was Approved by Army Corps Over Objections of Three Federal Agencies," *Inside Climate News,* August 30, 2016, https://insideclimatenews.org/news/30082016/dakota-access-pipeline-standing-rock-sioux-army-corps-engineers-approval-environment.

6. LaDonna Brave Bull Allard, "Why the Founder of Standing Rock Sioux Camp Can't Forget the Whitestone Massacre," *Yes! Magazine,* September 3, 2016. All quotations that follow are from this source.

7. Indigenous Environmental Network, "Indigenous Women Leaders of Dakota Access Pipeline Resistance to Speak Out For Protection of Earth and Water," *Common Dreams,* September 26, 2016, www.commondreams.org/newswire/2016/09/28/indigenous-women-leaders-dakota-access-pipeline-resistance-speak-out-protection.

8. Xian Chiang-Waren, "Inside the Camp That's Fighting to Stop the Dakota Access Pipeline," *Grist,* September 16, 2016, *http://grist.org/justice/inside-the-camp-thats-fighting-to-stop-the-dakota-access-pipeline/.*

9. International Indigenous Youth Council, September 3, 2016, www.nodaplarchive.com/international-indigenous-youth-council.html.

10. Mary Annette Pember, "Standing Ground on NoDAPL: Oceti Sakowin School Educates Next Generation," *Indian Country Today,* October 19, 2016, http://indiancountrytodaymedianetwork.com/2016/10/19/standing-ground-nodapl-oceti-sakowin-school-educates-next-generation-166137. See also Sheila Regan, "Amid Dakota Access Pipeline Protests, a Makeshift Native School Empowers Young Activists," *Fusion,* September10,2016,http://fusion.net/story/345619/dakota-access-pipeline-native-school-young-activists-defenders-of-the-water-school/.

11. Shelley Streeby, *Radical Sensations: World Movements, Violence, and Visual Culture* (Berkeley: University of California Press, 2013). See also David Graeber, *Direct Action: An Ethnography* (Oakland: A.K. Press, 2009).

12. Regan, "Amid Dakota Access Pipeline Protests"; Ann Arbor Miller and Dan Gunderson, "Photos: Winona's Kitchen Helps Feed Pipeline Protestors at Standing Rock," *MPR News,* https://www.mprnews.org/story/2016/11/23/photos-winonas-kitchen-helps-feed-pipeline-protesters-at-standing-rock.

13. "Cuba-Trained Doctors Head to Standing Rock," *Popular Resistance,* December 3, 2016, https://popularresistance.org/cuba-trained-doctors-head-to-standing-rock/.

14. Jack Healy, "From 280 Tribes, a Protest on the Plains," *New York Times,* September 12, 2016, www.nytimes.com/interactive/2016/09/12/us/12tribes.html.

15. Rich Kearns, "Solidarity from the South: Indigenous Leaders from Ecuador Come to Standing Rock," Indian Country Today, October 1, 2016, *http://indiancountrytodaymedianetwork.com/2016/10/01/solidarity-south-indigenous-leaders-ecuador-come-standing-rock-165962.*

16. Traci Voyles, *Wastelanding: Legacies of Uranium Mining in Navajo Country* (Minneapolis: University of Minnesota Press, 2015), 7.

17. Katherine Bagley, "At Standing Rock, a Battle over Fossil Fuels and Land," *Yale Environment 360,* November 10, 2016, http://e360.yale.edu/features/at_standing_rock_battle_over_fossil_fuels_and_land. All quotations that follow are from this source.

18. Edward Valandra, "We Are Blood Relatives: No to the DAPL," *Hot Spots, Cultural Anthropology* website, December 22, 2016,

https://culanth.org/fieldsights/1023-we-are-blood-relatives-no-to-the-dapl.

19. "Black Lives Matter Stands in Solidarity with Water Protectors at Standing Rock," http://blacklivesmatter.com/solidarity-with-standing-rock/. All quotations that follow are from this source.

20. NYC Stands with Standing Rock Collective, "#StandingRock-Syllabus," 2016, https://nycstandswithstandingrock.wordpress.com/standingrocksyllabus/. All quotations that follow are from this source.

21. Associated Press, "Oil Pipeline Protest Turns Violent in North Dakota," *NBC News,* September 4, 2016, www.nbcnews.com/storyline/dakota-pipeline-protests/oil-pipeline-protest-turns-violent-north-dakota-n642626. See also Sara Johnson, "'You Should Give Them Blacks to Eat': Cuban Bloodhounds and the Waging of an Inter-American War of Torture and Terror," *American Quarterly* 61, no. 1 (March 2009): 65–92.

22. Marlena Baldacci, Emanuella Grinberg, and Holly Yan, "Dakota Access Pipeline: Police Remove Protesters; Scores Arrested," *CNN,* October 27, 2016, www.cnn.com/2016/10/27/us/dakota-access-pipeline-protests/. Derek Hawkins, "Dakota Access Protesters Accuse Police of Putting Them in 'Dog Kennels,' Marking Them with Numbers," *Washington Post,* November 1, 2016, www.washingtonpost.com/news/morning-mix/wp/2016/11/01/dakota-access-protesters-accuse-police-of-putting-them-in-dog-kennels-marking-them-with-numbers/?utm_term = .e4bfe5fcfb72.

23. Sandy Tolan, "North Dakota Pipeline Activists Say Arrested Protesters Were Kept in Dog Kennels," *Los Angeles Times,* October 28, 2016, www.latimes.com/nation/la-na-north-dakota-pipeline-20161028-story.html.

24. Evelyn Nieves, "Police Attacked Standing Rock Activists for Hours," *Nation,* November 25, 2016, www.thenation.com/article/police-attacked-standing-rock-activists-for-hours-why-are-they-calling-it-a-riot/.

25. Sandy Tolan, "Thousands of Veterans Converge on North Dakota to Aid Pipeline Protest," *Los Angeles Times,* December 3, 2016, www.latimes.com/nation/la-na-dakota-access-protest-20161203-story.html.

26. "Thousands of Veterans Travel to Standing Rock to Support Activists," *NPR,* December 4, 2016, www.npr.org/2016/12/04/504352885/thousands-of-veterans-travel-to-standing-rock-to-support-activists.

27. Grace Dillon, *Walking the Clouds: An Anthology of Indigenous Science Fiction* (Tucson: University of Arizona Press, 2012), 3. Hereafter citations will appear in text. For other readings of Silko's work and *Almanac of the Dead* in relation to climate change activism, see Joni Adamson, "'¡Todos Somos Indios!' Revolutionary Imagination, Alternative Modernity, and Transnational Organizing in the Work of Silko, Tamez, and Anzaldúa," *Journal of Transnational American Studies* 4, no. 1 (2012): 1–26; Shari Huhndorf, *Mapping the Americas: The Transnational Politics of Contemporary Native Culture* (Ithaca, NY: Cornell University Press, 2009); Sarah Ray, "Environmental Justice, Transnationalism, and the Politics of the Local in Leslie Marmon Silko's *Almanac of the Dead,*" *Journal of Transnational American Studies* 5, no. 1 (2013): 1–24; T.V. Reed, "Toxic Colonialism, Environmental Justice, and Native Resistance in Silko's *Almanac of the Dead,*" MELUS 34, no. 2 (2009): 25–42; and Claudia Sadowski-Smith, *Border Fictions: Globalization, Empire, and Writing at the Boundaries of the United States* (Charlottesville: University of Virginia Press, 2008). See also Mishuana Goeman, *Mark My Words: Native Women Mapping Our Nations* (Minneapolis: University of Minnesota Press, 2013).

28. Gerald Vizenor, *Bearheart: The Heirship Chronicles* (Minneapolis: University of Minnesota Press, 1990). Hereafter citations appear in text.

29. Thomas Banyacya, "The Hopi Prophecy," in *Surviving in Two Worlds: Contemporary Native American Voices,* edited by Lois Crozier-Hogle and Darryl Babe Wilson (Austin: University of Texas Press, 1997), 40–51.

30. Jack Loeffler, "Black Mesa (New Mexico)," in *Encyclopedia of Religion and Nature,* edited by Bron Taylor (New York: Continuum International, 2005), 200.

31. "Poison Fire–Sacred Earth," *Testimonies, Lectures, Conclusions: The World Uranium Hearing, Salzburg 1992* (Munchen, Germany: 1993), 32–36. All quotations that follow are from this source. Excerpts from Bancacya's testimony cited in the body of this book can also be accessed at *https://ratical.org/radiation/WorldUraniumHearing/ThomasBanyacya.html.*

32. "Letter to David Chatfield, Board Chairman," Greenpeace USA, May 24, 1990, www.ejnet.org/ej/SNEEJ1.pdf.

33. Southwest Organizing Project, "Letter to Big 10 Environmental Groups," March 1990, www.ejnet.org/ej/swop.pdf.

34. "Declaration of Quito," *Native Web,* July 1990, www.nativeweb.org/papers/statements/quincentennial/quito.php. All quotations that follow are from this source.

35. Dana Alston, "The Summit: Transforming a Movement," in "The Summit," special issue, *Race, Poverty, and the Environment* 2, no. 3/4 (1991). Reprinted in *Race, Poverty, and the Environment* 17, no. 1 (Spring 2010): 14–17.

36. "About," *Indigenous Environmental Network,* www.ienearth.org/about/. Accessed 24 July 2017.

37. E.A. Parson, P.M. Haas, and M.A. Levy, "A Summary of Major Documents Signed at the Earth Summit and the Global Forum," *Environment* 34, no. 4 (1992): 12–15, 34–36.

38. "Report of the UN Conference on Environment and Development," Rio de Janeiro, June 3–14, 1992, www.un.org/documents/ga/conf151/aconf15126–1annex1.htm.

39. Luke Cole and Sheila Foster, *From the Ground Up: Environmental Racism and the Rise of the Environmental Justice Movement* (New York: New York University Press, 2001), 141. All quotations that follow are from this source.

40. Zoltan Grossman and Al Gedicks, "Native Resistance to Multinational Mining Corporations in Wisconsin," *Cultural Survival Quarterly,* March 2001, *www.culturalsurvival.org/publications/cultural-survival-quarterly/native-resistance-multinational-mining-corporations.*

41. William Lempert, "Indigenous Filmmakers Reimagine Science," *Journal of Labocene,* 29 January 2017, *https://medium.com/labocine/indigenous-filmmakers-reimagine-science-4e0690183924.* See also Lempert, "Decolonizing Encounters of the Third Kind: Alternative Futuring in Native Science Fiction Film," *Visual Anthropology Review* 30, no. 2 (2014): 164–76 and "Navajos on Mars: Native Sci-Fi Film Futures," *Medium,* 21 September 2015, *https://medium.com/space-anthropology/navajos-on-mars-4c336175d945.* See also Zoltan Grossman, "Māori Opposition to Fossil Fuel Development in

Aotearoa New Zealand," available at *http://academic.evergreen.edu/g/grossmaz/MaoriOilOppositionArticle.pdf* and Terrence Loomis, *Petroleum Development and Environmental Conflict in Aotearoa New Zealand: Texas of the South Pacific* (New York: Lexington Books, 2016).

CHAPTER TWO. CLIMATE REFUGEES IN THE GREENHOUSE WORLD

1. OEB 3245, commonplace book (large), Octavia E. Butler Papers, The Huntington Library, San Marino, California, 4. Hereafter, citations follow the Huntington's system.

2. OEB 3248, commonplace book (large).

3. Donna Haraway, *Staying with the Trouble: Making Kin in the Chthulucene* (Durham, NC: Duke University Press, 2016), 62–63.

4. OEB 3245, commonplace book (large).

5. OEB 3245, commonplace book (large).

6. OEB 3221, commonplace book (large).

7. Radio Imagination Press Release, https://clockshop.org/project/radio-imagination/. Recent scholarship drawing on Butler's archive includes Aimee Bahng, "Plasmodial Improprieties: Octavia Butler, Slime Molds, and Imagining a Femi-Queer Commons," *Queer Feminist Science Studies Reader* (Seattle: University of Washington Press: forthcoming); Gerry Canavan, *Modern Masters of Science Fiction: Octavia E. Butler* (Urbana: University of Illinois, 2016); Kim Hester-Williams, "Earthseeds of Change: Post-Apocalyptic Mythmaking, Race, and Ecology in the *Book of Eli* and Octavia Butler's Womanist Parables," in *Racial Ecologies*, edited by Hester-Williams and LeiLani Nishime (Seattle: University of Washington Press, forthcoming 2018); Ayana Jamieson, "Black Blessings: Toni Cade Bambara and Octavia E. Butler," *Feminist Wire*, November 23, 2014, www.thefeministwire.com/2014/11/black-women-writers/, and Butler's biography, which Jamieson is currently writing; and Sami Schalk, *Bodyminds Reimagined: (Dis)ability, Race, and Gender in Black Women's Speculative Fiction* (Durham, NC: Duke University Press, forthcoming 2018). Alexis Lothian, *Old Futures: Speculative Fiction and Queer Possibility* (New York: New York

University Press, forthcoming 2018) and Aimee Bahng, *Migrant Futures: Decolonizing Speculation in Financial Times* (Durham, NC: Duke University Press, 2017) have also crucially shaped my thinking here and throughout.

8. Butler's pathway through different educational institutions and experiences and her endless archiving recall Fred Moten's and Stefano Harney's students in *The Undercommons: Fugitive Planning and Black Study* (Oakland: AK Press, 2013) who are "committed to black study in the university's undercommon rooms" and who "study without an end, plan without a pause, rebel without a policy, conserve without a patrimony," who "never graduate. They just ain't ready. They're building something in there, something down there" (67).

9. David Harvey, *A Brief History of Neoliberalism* (New York: Oxford University Press, 2005), 2.

10. OEB 3263, commonplace book (large).

11. OEB 3823–24, Octavia E. Butler to Octavia M. Butler.

12. OEB 4176, Octavia E. Butler to Marjorie Rae Nadler.

13. OEB 297, Clarion character sketches: notes and fragments.

14. OEB 3794, Octavia E. Butler to Russell Bates.

15. Butler Papers, box 284: Subject files—Places cont'd (1) Latin America. ca. 1977, Octavia E. Butler Papers, The Huntington Library, San Marino, California.

16. Box 281: Subject files—Minorities (1), Octavia E. Butler Papers.

17. OEB 1001, journal, May 7, 1976.

18. Walidah Imarisha, "Introduction," in *Octavia's Brood: Science Fiction Stories from Social Justice Movements,* edited by adrienne maree brown and Walidah Imarisha (Oakland: AK Press, 2015), 4.

19. OEB 3242, commonplace book (large).

20. OEB 1772, Parable of the Sower: novel: outline.

21. OEB 3234, commonplace book (large).

22. OEB 3242, commonplace book (large). See also Canavan, *Modern Masters of Science Fiction,* 123–25.

23. OEB 3240, commonplace book (large).

24. OEB 2082, Parable of the Trickster: novel: notes and fragments.

25. OEB 3240, commonplace book (large).
26. OEB 2135, Parable of the Trickster: novel: fragment.
27. OEB 3242, commonplace book (large).
28. OEB 3243, commonplace book (large).
29. OEB 3248, commonplace book (large).
30. See also Canavan, *Modern Masters of Science Fiction*, 139–40.
31. OEB 2041, Parable of the Trickster: novel: notes.
32. OEB 2049: Parable of the Trickster: novel: notes.
33. OEB 3279, commonplace book (large).
34. OEB 2075, Parable of the Trickster: novel: yellow binder: notes.
35. OEB 2055, Parable of the Trickster: novel: notes,
36. OEB 2076, Parable of the Trickster: novel: yellow binder: notes.
37. OEB 2055, Parable of the Trickster: novel: notes.
38. OEB 2076, Parable of the Trickster: novel: yellow binder: notes.
39. OEB 2055, Parable of the Trickster: novel: notes.
40. OEB 2124, Parable of the Trickster: novel: notes.
41. Box 288: Subject files—Disaster; The environment; TV, film, radio, theater; box 295: Research materials—Social conditions; Disaster; The environment; Box 348: Research materials—Extra oversize; and box 349: Research materials—Extra oversize, Octavia E. Butler Papers.
42. OEB 94, author's questionnaire.
43. "Scientists See Change in the Weather," *Los Angeles Times*, January 22, 1977, p. 27.
44. Harriet Stix, "'Solar Energy Easier to Sell': Lower Power Bills an Inducement, Builder Says," *Los Angeles Times*, February 6, 1977.
45. Patrick Boyle, "California Contrast: Drought and a New Lake," *Los Angeles Times*, February 6, 1977.
46. Barry Commoner, "Solar Energy Could Avert Crisis in Southern California," *Los Angeles Times*, February 6, 1977.
47. John A. Jones, "More than a Ray of Hope: Using Sun as Alternative to Oil, Gas No Longer Far-Out Idea," *Los Angeles Times*, April 8, 1978, p. 17.
48. S.J. Diamond, "Reports on Ecology: A New Industry," *Los Angeles Times*, January 6, 1977; Robert Gillette, "Drastic Warming of Climate Feared: Study Warns of Reliance on Coal, Oil," *Los Angeles Times*, July 25, 1977, p. 16.

49. OEB 3176, Commonplace books (medium).
50. Box 288, (5) The environment, Octavia E. Butler Papers.
51. Box 295, (6) The environment (1), Octavia E. Butler Papers.
52. Box 295, (7) The environment (2), Octavia E. Butler Papers.
53. Box 277, (3) Science, Octavia E. Butler Papers.
54. Box 295, (6) The environment (1), Octavia E. Butler Papers.
55. Box 349, (3) Research materials: The environment, Octavia E. Butler Papers.
56. Box 295, (7), The environment (2), Octavia E. Butler Papers.
57. Box 295, (6) The environment, Octavia E. Butler Papers.
58. OEB 1702, Parable of the Sower: novel: notes.
59. OEB 3238, commonplace book (large).
60. OEB 1708, Parable of the Sower: novel: notes.
61. Rob Nixon, *Slow Violence and the Environmentalism of the Poor* (Cambridge, MA: Harvard University Press, 2013).
62. OEB 3080, Parable of the Sower: speech: notecards.
63. OEB 1725–53, Parable of the Sower: novel: fragments.
64. OEB 1726, Parable of the Sower: early fragment.
65. OEB 1735, Parable of the Sower: novel: early fragment.
66. Rick Noack, "Has the Era of 'Climate Change' Refugee Begun?" *Washington Post,* August 7, 2014.
67. OEB 3248, commonplace book (large).
68. OEB 3265, commonplace book (large).
69. OEB 3265, commonplace book (large).
70. OEB 1065, journal.

CHAPTER THREE. CLIMATE CHANGE AS A WORLD PROBLEM

1. Adrienne maree brown, *Emergent Strategy: Shaping Change, Shaping Worlds* (Oakland: AK Press, 2017), 58. Hereafter page numbers appear parenthetically in text.
2. "Detroit Narrative Agency," Allied Media Projects, www.alliedmedia.org/dna, accessed 24 July 2017.
3. See "Detroit Future," www.alliedmedia.org/detroit-future/story, accessed 24 July 2017; Tawana Petty, "Detroiters Demystify

'Data' through DiscoTechs," October 25, 2016, *Detroit Digital Justice Coalition,* http://detroitdjc.org; "Black-Mesa-Water-Coalition" Facebook Page, https://m.facebook.com/pg/blackmesawc/about/?entry_point = page_nav_about_item&ref = page_internal, accessed 24 July 2017; "About Incite!" www.incite-national.org/page/about-incite, accessed 24 July 2017.

4. Byrd-Hagel Resolution, 105th Congress, 1st Session, S. Res. 98, July 25, 1997, www.congress.gov/bill/105th-congress/senate-resolution/98?r = 3.

5. Ed King, "Kyoto Protocol: 10 Years of the World's First Climate Change Treaty," *Climate Change News,* November 16, 2015, www.climatechangenews.com/2015/02/16/kyoto-protocol-10-years-of-the-worlds-first-climate-change-treaty/.

6. David Biello, "U.S., China, India and Other Nations Arrive at Nonbinding Agreement at U.N. Climate Summit," *Scientific American,* December 18, 2009, www.scientificamerican.com/article/us-china-india-climate-accord/.

7. "Negotiators Fail to Deliver at Doha Climate Summit: Outcome is Nowhere Near What Is Needed to Meet 2°C Target," December 8, 2012, *Union of Concerned Scientists,* www.ucsusa.org/news/press_release/negotiators-spew-more-hot-air-in-doha-0353.html#.WGcPQLGZPwc.

8. "What Is RT?" *https://risingtidenorthamerica.org/features/what-is-rising-tide/*, accessed 24 July 2017

9. Frederika Whitehead, "The First Climate Justice Summit: A Pie in the Face for the Global North," *Guardian,* April 16, 2014, www.theguardian.com/global-development-professionals-network/2014/apr/16/climate-change-justice-summit.

10. "Bali Principles of Climate Justice," International Climate Justice Network, August 28, 2002, available at *EJNetOrg,* www.ejnet.org/ej/bali.pdf.

11. Manuel Pastor, Robert Doyle Bullard, James K. Boyce, Alice Fothergill, Rachel Morello-Frosch, Beverly Wright, *In the Wake of the Storm: Environment, Disaster and Race after Katrina* (New York: Russell Sage Foundation: 2006). See also Rebecca Solnit, *A Paradise Built in Hell: The Extraordinary Communities that Arise in Disaster* (New York: Penguin, 2009).

12. "Inuit Petition: Inter-American Commission on Human Rights to Oppose Climate Change Caused by the United States of America," Inuit Circumpolar Council, December 7, 2005, www.inuitcircumpolar.com/inuit-petition-inter-american-commission-on-human-rights-to-oppose-climate-change-caused-by-the-united-states-of-america.html.

13. Andrew C. Revkin, "Inuit Climate Change Petition Rejected," *New York Times,* December 16, 2006, www.nytimes.com/2006/12/16/world/americas/16briefs-inuitcomplaint.html.

14. "U.N. Declaration on the Rights of Indigenous Peoples," March 2008, www.un.org/esa/socdev/unpfii/documents/DRIPS_en.pdf.

15. "Inuit Call on Global Leaders at CoP 16: Help us Sustain our Homeland: Take Immediate Action on Climate Change," November 29, 2010, www.inuitcircumpolar.com/press-releases-2010.html.

16. Leyland Cecco, "The Face of Climate Change: How Inuit Youth Lead the Fight to Save the Arctic," *Guardian,* December 10, 2015, www.theguardian.com/environment/2015/dec/10/paris-climate-change-talks-arctic-inuit-youth-cop21.

17. Inuit Circumpolar Panel Council 2015–2016 Report, www.inuitcircumpolar.com/uploads/3/0/5/4/30542564/english_final_2015–2016_annual_report_withcover_fordigital.pdf.

18. Adrienne maree brown, "Oh Canada," May 16, 2006, on her personal website, http://adriennemareebrown.net/2006/05/16/oh-canada/.

19. Adrienne maree brown, "The Green Generation," in *The Global Warming Reader: A Century of Writing About Climate Change,* edited by Bill McKibben, 260 (New York: Penguin, 2012). Hereafter page numbers appear in text.

20. See the film *American Revolutionary: The Evolution of Grace Lee Boggs* (dir. Lee, 2014).

21. Julia Putnam, "A Lifelong Search for Real Education," *Yes! Magazine,* October 16, 2015, www.yesmagazine.org/issues/learn-as-you-go/a-lifelong-search-for-real-education. See also Grace Lee Boggs with Scott Kurashige, *The Next American Revolution: Sustainable Activism for the Twenty-First Century* (Berkeley: University of California Press, 2012).

22. "RUCKUS Society," on their website, http://ruckus.org.

23. "Interview with Adrienne Maree Brown: Voices of Climate Justice," *Climate Change: Catalyst or Catastrophe?* 16, no. 2 (Fall 2009), www.reimaginerpe.org/cj/brown.

24. "Interview with Adrienne Maree Brown," *Climate Change.*

25. Moya Bailey, "'Shaping God': The Power of Octavia Butler's Black Feminist and Womanist SciFi Visions in the Shaping of a New World: An Interview with Adrienne Maree Brown," *ADA: A Journal of Gender, New Media, and Technology* 3 (2013), http://adanewmedia.org/2013/11/issue3-baileybrown/.

26. Adrienne maree brown, "Outro," in *Octavia's Brood: Science Fiction Stories from Social Justice Movements,* edited by adrienne maree brown, 279 (Oakland: AK Press, 2015). Hereafter page numbers appear in text.

27. Brown, "Hiatus," November 8, 2014, on her personal website, http://adriennemareebrown.net/2014/11/08/hiatus/.

28. Dani McClain, "Homing Instinct," in *Octavia's Brood: Science Fiction Stories from Social Justice Movements,* edited by adrienne maree brown and Walidah Imarisha, 539 (Oakland: AK Press, 2015). Hereafter citations appear in text.

29. Korematsu was a Japanese American born in Oakland who refused to comply with Franklin D. Roosevelt's Executive Order 9066 mandating that West Coast people of Japanese descent be removed from their homes and forced to live in internment camps. Shakur is a former member of the Black Panther Party who escaped from prison in 1979 and fled to Cuba in 1983, where she was given asylum and has lived ever since.

30. Adrienne maree brown, "The River," in *Octavia's Brood: Science Fiction Stories from Social Justice Movements,* edited by adrienne maree brown and Walidah Imarisha, 23 (Oakland: AK Press, 2015). Hereafter citations appear in text.

31. Grace Lee Boggs, "Youth and Social Justice in Education," in *Handbook of Social Justice in Education,* edited by William Ayers, Therese M. Quinn, David Stovall, 522 (London: Routledge, 2009).

32. Adrienne maree brown, "Don't Patronize Us/Them (From Oakland to Standing Rock)," December 5, 2016, on her personal website,http://adriennemareebrown.net/2016/12/05/dont-patronize-usthem-from-oakland-to-standing-rock/.

GLOSSARY

DIRECT ACTION Direct action may take such forms as protests, sit-ins, blockades, boycotts, and hacktivism and is an important tactic for many social movements wary of making the state the horizon of possibility. It has its roots in anticolonial, antislavery, and labor struggles that extend backwards in time for centuries. In the 1910s, the Industrial Workers of the World made it central to their radical world-making. It was a keyword for Martin Luther King, Jr., and for the Black freedom struggles of the 1960s as well as for antiwar and environmental movements ever since. It was also a key tactic for the American Indian Movement and the American Indian Youth Council. The Standing Rock Youth Council takes "non-Violent Direct Action" to advance their "voices in decisions made about the future of Indian Country."

SPECULATIVE FICTION Speculative fiction is an umbrella term encompassing science fiction, fantasy, horror, cli-fi, and related genres. In the late nineteenth century, literary critics used it to classify utopian fiction such as Edward Bellamy's *Looking Backward: 2000–1887*. In the mid-twentieth century, writer Robert Heinlein expressed a preference for the term and contrasted it with fantasy in order to provide a more respectable intellectual genealogy for the genre. At the turn of the twenty-first century, Sheree Thomas chose it over "science fiction" in her influential anthology *Dark Matter: Speculative Fiction from the African Diaspora* (2000) as a way of broadening ideas about what counted as part of the history of the genre. In this book I use both terms but often choose speculative fiction to recognize more heterogeneous kinds of speculative writing and cultural production and more expansive understandings of science and technology.

SYMBIOSIS The Greek roots of the word refer to living together, companionship, and partnering. In biology, since at least 1882 according to the Oxford English Dictionary, it has signified "the association of two different organisms (usually two plants, or an animal and a plant) which live attached to each other, or one as a tenant of the other, and contribute to each other's support. Also more widely, any intimate association of two or more different organisms, whether mutually beneficial or not." The authors of "Let Our Indigenous Voices Be Heard" envision a productive symbiosis, based upon mutual respect, between Indigenous and Western knowledges that could serve shared goals of sustainability in the face of climate change. Octavia E. Butler imagines symbiotic entanglements among humans, critters, and the earth that belie myths of isolated, competitive individuals. Adrienne maree brown partners with communities and movements, using direct action to confront climate change and environmental racism and co-create "symbiotic relationships based on our needs and our dreams."

WORLD-MAKING World-making, as Alexis Lothian explains, has long been an important term for queer theorists and writers who imagine queer practices as transformative portals to futurity and potentiality. She contrasts queer world-making with the science fiction term world-building, which refers to the creation of plausible worlds, even as she recognizes that some science fiction worlds break with the present to posit a future whose difference makes a case for political change. World-building is often associated with colonization and empire though it also may be used to critically interrogate those structures. In this book I choose world-making over world-building to center the transformative dimensions of the worlds and futures imagined by Indigenous people and people of color in confronting settler colonialism, environmental racism, and climate change.

KEY FIGURES

LADONNA BRAVE BULL ALLARD Lakota historian who founded the Sacred Stone resistance camp on her property when she heard the Energy Transfer Partners pipeline would be routed near her water well and her son's grave.

SVANTE ARRHENIUS Swedish scientist who in 1896 was the first to ask what effects increasing atmospheric carbon dioxide would have on the world's climate.

THOMAS BANYACYA Hopi leader charged by elders after World War II with bringing the Hopi message about the future to the English-speaking world. He spoke to UN assemblies and other international bodies, condemning the strip-mining of Black Mesa and other environmental damage humans were doing to the Earth.

BONG JOON-HO South Korean director of *Snowpiercer* (2013), who says that the genre of science fiction allowed him to ask big questions about capitalism, power, geo-engineering, and climate change.

ADRIENNE MAREE BROWN Detroit social movement organizer, theorist, and speculative fiction writer who uses the work of Octavia E. Butler in what she calls intersectional emergent strategy.

OCTAVIA E. BUTLER One of the greatest science fiction writers of all time, author of the classic climate change novel *Parable of the Sower*

(1993), and HistoFuturist organizer of a massive archive of material on climate change, science, the environment, disaster, and neoliberalism from the 1960s through the mid-2000s.

RACHEL CARSON Aquatic biologist who became a nature writer in the 1950s and authored *Silent Spring* (1962), which partly relied on speculative fiction to change the world by inspiring the modern environmental movement.

JOSEPH FOURIER French mathematician and physicist credited with discovering in the 1820s what is now called the greenhouse effect, the process by which some of the sun's energy that reaches the Earth is reflected back into space and the rest is absorbed and returned to the atmosphere as heat.

JAMES HANSEN Climate scientist whose 1988 Congressional testimony that global warming was already happening helped call attention to the problem of climate change.

WALIDAH IMARISHA Co-editor, with adrienne maree brown, of *Octavia's Brood: Science Fiction Stories from Social Justice Movements* (2015), theorist of visionary fiction, and author of "Black Angel" and other works of fiction and nonfiction.

CHARLES DAVID KEELING Scientist whose collection of carbon dioxide samples on Mauna Loa, Hawai'i, from 1958 to 2005 showed that there was an increase in carbon dioxide over time. This rise in the carbon dioxide in the atmosphere is now called the Keeling Curve.

RONALD REAGAN Elected U.S. president in 1980 and serving two terms, Reagan was hostile to the small gains that the emerging environmental movement had made, pushed deregulation, and prioritized untrammeled economic growth over confronting environmental harms.

ROGER REVELLE Scientist at Scripps Institute of Oceanography in La Jolla, California, who in 1957 suggested that oceans would not act as sinks for human-produced carbon dioxide to the extent previously thought and that greenhouse gases would eventually produce global warming.

LESLIE MARMON SILKO Laguna Pueblo writer and author of *Almanac of the Dead* (1991), an early novel that imagined possibilities that might emerge in the wake of climate change disaster.

JULES VERNE French writer who is a crucial figure in the early history of science fiction and author of *The Purchase of the North Pole* (1889), a novel about a geo-engineering scheme in the Arctic.

GERALD VIZENOR Anishinaabe writer and scholar who used the term "slipstream" as early as the 1970s and during that decade also wrote *Darkness in Saint Louis: Bearheart* (1978), a speculative fiction about resource extraction on Indigenous lands.

JAMES WATT Secretary of the Interior in the Reagan Administration from 1981 to 1983, who was hostile to environmentalism and opened up public lands for resource extraction and development.

SELECTED BIBLIOGRAPHY

Adamson, Joni. *American Indian Literature, Environmental Justice, and Ecocriticism: The Middle Place.* Tucson: University of Arizona Press, 2001.

Adamson, Joni, and Michael Davis, eds. *Humanities for the Environment: Integrating Knowledges, Forging New Constellations of Practice.* London: Routledge, 2017.

Adamson, Joni, Mei Mei Evans, Rachel Stein, eds. *The Environmental Justice Reader: Politics, Poetics, and Pedagogy.* Tucson: University of Arizona Press, 2002.

Alaimo, Stacy. *Exposed: Environmental Politics and Pleasures in Posthuman Times.* Minneapolis: University of Minnesota Press, 2016.

Bahng, Aimee. *Migrant Futures: Decolonizing Speculation in Financial Times.* Durham, NC: Duke University Press, 2017.

Boggs, Grace Lee, with Scott Kurashige. *The Next American Revolution: Sustainable Activism for the Twenty-First Century.* Berkeley: University of California Press, 2012.

brown, adrienne maree. *Emergent Strategy: Shaping Change, Shaping Worlds.* Oakland: AK Press, 2017.

brown, adrienne maree, and Walidah Imarisha, eds. *Octavia's Brood: Science Fiction Stories from Social Justice Movements.* Oakland: AK Press, 2015.

Butler, Octavia E. Octavia E. Butler Papers. The Huntington Library, San Marino, California.

———. *Parable of the Sower.* New York: Grand Central Publishing, 2000; 1993.

———. *Parable of the Talents.* New York: Seven Stories Press, 1998.

Carson, Rachel. *Silent Spring.* New York: Houghton Mifflin, 2002; 1962.

Dillon, Grace. *Walking the Clouds: An Anthology of Indigenous Science Fiction.* Tucson: University of Arizona Press, 2012.

Ghosh, Amitav. *The Great Derangement: Climate Change and the Unthinkable.* Chicago: University of Chicago Press, 2016.

Graeber, David. *Direct Action: An Ethnography.* Oakland: AK Press, 2009.

Hamilton, Clive. *Earthmasters: The Dawn of the Age of Climate Engineering.* New Haven: Yale University Press, 2013.

Haraway, Donna. *Staying with the Trouble: Making Kin in the Chthulucene.* Durham, NC: Duke University Press, 2016.

Harvey, David. *A Brief History of Neoliberalism.* New York: Oxford University Press, 2005.

Klein, Naomi. *This Changes Everything: Capitalism vs. the Climate.* New York: Simon and Schuster, 2014.

LaDuke, Winona. *The Winona LaDuke Chronicles: Stories from the Front Lines in the Battle for Environmental Justice.* Winnipeg: Fernwood Publishing, 2017.

LeMenager, Stephanie. *Living Oil: Petroleum Culture in the American Century.* London: Oxford, 2014.

Lothian, Alexis. *Old Futures: Speculative Fiction and Queer Possibility.* New York: New York University Press, forthcoming 2018.

Lovelock, James. *Gaia: A New Look at Life on Earth.* New York: Oxford, 1979.

McKibben, Bill, ed. *The Global Warming Reader: A Century of Writing about Climate Change.* New York: Penguin, 2012.

Nixon, Rob. *Slow Violence and the Environmentalism of the Poor.* Cambridge, MA: Harvard University Press, 2011.

NYC Stands with Standing Rock Collective. "#StandingRock Syllabus." 2016. https://nycstandswithstandingrock.wordpress.com/standingrocksyllabus/.

Pellow, David N. *Resisting Global Toxics: Transnational Movements for Environmental Justice.* Cambridge, MA: MIT Press, 2007.

Pulido, Laura. *Environmentalism and Economic Justice: Two Chicano Struggles in the Southwest.* Tucson: University of Arizona Press, 1996.

Silko, Leslie Marmon. *Almanac of the Dead.* New York: Penguin, 1992.

Siperstein, Stephen, Shane Hall, and Stephanie LeMenager, eds. *Teaching Climate Change in the Humanities.* New York: Routledge, 2017.

Solnit, Rebecca. *A Paradise Built in Hell: The Extraordinary Communities That Arise in Disaster.* New York: Penguin, 2009.

Stephenson, Wen. *What We're Fighting for Now Is Each Other: Dispatches from the Front Lines of Climate Justice.* Boston: Beacon Press, 2015.

Voyles, Traci. *Wastelanding: Legacies of Uranium Mining in Navajo Country.* Minneapolis: University of Minnesota Press, 2015.

Weaver, Jace, ed. *Defending Mother Earth: Native American Perspectives on Environmental Justice.* Maryknoll, NY: Orbis Books, 1996.